Comments on Books by Michael J. Crowe

Michael J. Crowe is the author of six previous books. Listed below are published comments by other scholars on some of these books.

A History of Vector Analysis: The Evolution of the Idea of a Vectorial System. This book won a Jean Scott prize from La Maison des Sciences de l'Homme in Paris.

T. A. A. Broadbent (*Journal of the Franklin Institute*, 1971): "Professor Crowe tells [his story] clearly, in full detail, with excellent documentation. . . . a fascinating volume, at once informative and readable."

Howard Eves (*Science*, 1969) described it as "scholarly and painstaking. Crowe tells his story completely and magnificently—sometimes in almost majestically structured sentences."

Theories of the World from Antiquity to the Copernican Revolution

David Hughes (*Observatory*, 1991): "I recommend this book to you without reservation. [It] makes the history of astronomy really exciting."

Robert Hatch (*Isis*, 1991): "Deftly survey[s] astronomical ideas, arguments, and techniques from classical Greece to the early decades of the seventeenth century."

William H. Donahue (*American Scientist*, 1992): "*Theories of the World* accomplishes its worthy purpose competently and engagingly. [It] deserves wide distribution."

The Extraterrestrial Life Debate 1750–1900: The Idea of a Plurality of Worlds from Kant to Lowell, which is also available in a three volume Japanese translation.

David Hughes (*New Scientist*, 1986): "There isn't an uninteresting page in [this] long text. It is a masterly review of an intriguing subject, erudite and entertaining, clear and all-encompassing."

E. Robert Paul (*Sky and Telescope*, 1987): "Crowe's book is lucid and rich in historical detail. His analysis is so fascinating and his comments on the contemporary debate so pertinent that [his book] can be recommended for the thoughtful reader without reservation. While a model of scholarly analysis, it has the unusual virtue of reading with the excitement of high adventure."

John Brooke (*Annals of Science*, 1987) described it as "fascinating and richly rewarding. A great strength of the book is the author's sensitivity to the religious issues. . . . fastidious research [producing] countless surprises."

Modern Theories of the Universe from Herschel to Hubble

Allan Chapman (*Annals of Science*, 1996) described it as "a clear and well-guided route through the fundamental concepts of cosmology over the past few centuries. [It] succeeds admirably in achieving its [stated] goal."

Calendar of the Correspondence of Sir John Herschel, edited by Crowe with David R. Dyck and James Kevin (associate editors).

Sydney Ross (*Isis*, 1999) described the book as "an unparalleled historical source that will be a goldmine for historians of nineteenth-century science from now on. The editors are to be congratulated on having carried out a task of heroic magnitude and wide usefulness."

Michael Hoskin (*Annals of Science*, 2000) describes it as a "splendid volume [that makes Herschel's correspondence] available with a convenience of which past researchers could only dream."

Brian Warner (*Observatory*, 1999): "The Calendar will be the starting point of all future Herschel scholarship."

Unsigned review (*European Astronomy*, 2002): "This volume is far and away the most extensive source of information on John Herschel ever published."

Mechanics from Aristotle to Einstein

Mechanics from Aristotle to Einstein

Michael J. Crowe

Green Lion Press

Santa Fe, New Mexico

Manufactured in the United States of America.

Published by Green Lion Press, Santa Fe, New Mexico

www.greenlion.com

Green Lion Press books are printed on fine quality acid-free paper of high opacity. Both softbound and clothbound editions feature bindings sewn in signatures. Sewn signature binding allows our books to open securely and lie flat. Pages do not loosen or fall out and bindings do not split under heavy use by students and researchers. Clothbound editions meet the guidelines for permanence and durability of the Committee on Production Guidelines for Book Longevity of the Council on Library Resources. The paper used in all Green Lion Press books meets the minimum requirements of American National Standard for Information Sciences— Permanence of Paper for Printed Library Materials, ANSI Z39.48-1984.

Set in 10-point Helvetica and 11-point Palatino. Printed and bound by Sheridan Books, Chelsea, Michigan.

Cover illustration by William H. Donahue based on an engraving in Isaac Newton, *System of the World* (1728).

Cataloging-in-publication data:
Crowe, Michael J.
Mechanics from Aristotle to Einstein / by Michael J. Crowe

Includes index, bibliography, biographical notes, extensive selections of original texts.

ISBN-13: 978-1-888009-32-3 (sewn softcover binding)

1. Mechanics. 2. History of Science. 3. Physics. 4. Aristotle. 5. Galileo. 6. Newton. 7. Einstein.

I. Michael J. Crowe (1936–). II. Title

Library of Congress Control Number 2007925664

Dedicated to four deeply inspiring teachers:

Harold E. Petersen,
who at Rockhurst High School introduced me
to the joys of the intellect,

Frederick J. Crosson and Willis D. Nutting,
who first as my teachers, then as colleagues at the
University of Notre Dame
encouraged my commitment
to a life of learning and teaching, and

Erwin N. Hiebert,
who at the University of Wisconsin inspired and directed
my doctoral studies and research.

Contents

The Green Lion's Preface

Green Lion Press is delighted to have been chosen by Michael J. Crowe to publish this book. Crowe invites readers to join him in the adventure of thinking through original works of Aristotle, Oresme, Galileo, Descartes, Newton, and others. We experience first-hand the genesis of the brilliantly successful science of moving bodies that led the way to the western scientific approach to the natural world.

The story that emerges is thrilling and surprisingly rich. The science of motion, far from being a dry and technical topic, involves philosophy, mathematics, theology, and even poetry and music, as well as ingenious experimentation and physical speculation. We encounter not only the careful development of recurrent themes, but also surprising reversals, as each succeeding leap of thought involves the rejection of truths once dearly held. The reader who follows it through to the end will come away with a new and deeper appreciation of the achievement that this extraordinary series of texts represents.

Mechanics from Aristotle to Einstein is thus a worthy companion to Professor Crowe's other guides and source books, which have proved their worth many times to a wide readership.

On our part, we have endeavored to make the format and construction of this book worthy of its contents.

Construction. The book is printed on acid-free paper and has a sewn, rather than glued, binding so as to survive years of heavy use and careful study.

Diagrams are so placed as to avoid the need to turn a page to follow the accompanying argument.

Fonts have been chosen so as to allow readers to distinguish immediately between the original source texts and Professor Crowe's supporting material. Crowe's text is in the current, more modern typeface (Helvetica), while the extended selections from original sources are in an earlier style (Palatino). Because of the distracting appearance of frequent font changes, shorter quotations from original sources are not set in Palatino, but are either included in the main text (when very brief) or set off by indentation and extra spacing. Sources of all such quotations are clearly identified.

Notes are included as footnotes, not end notes (as is increasingly, and lamentably, common in scholarly books). There are three sources for these notes: the author of the book (Crowe), the author of the original selections, and the translator or editor of the selections. Original author notes and translator's or editor's notes are identified as such in square brackets following the note. **Running heads** are designed to allow the reader to locate a particular passage readily.

We would like especially to commend the **Selected Bibliography** to the reader's attention. This bibiliography was prepared by Professor Crowe on the basis of many years' experience with the original sources and with the often daunting secondary literature on them. It is intended to give readers advice as to where to turn next for further study of one or another topic in the book. The bibliography is thus a well-informed and extremely valuable guide to deeper aspects of the history of mechanics.

We are confident that readers will find this book a worthy companion and guide on the adventure upon which they are embarking.

Dana Densmore and William H. Donahue
For the Green Lion Press

Preface

It is my conviction that the story of the development of mechanics from Galileo to Einstein's special theory of relativity is the most remarkable story in all secular history. The famous philosopher Alfred North Whitehead, writing in his *Science and the Modern World* about only that portion of the story from Galileo through Newton, suggested one reason why the story seems so noteworthy.

> This subject of the formation of the three laws of motion and of the law of gravitation deserves critical attention. The whole development of thought occupied exactly two generations. It commenced with Galileo and ended with Newton's *Principia*; and Newton was born in the year that Galileo died. Also the lives of Descartes and Huygens fell within the period occupied by these great terminal figures. The issue of the combined labors of these four men has some right to be considered as the greatest single intellectual success which mankind has achieved.[1]

Strong as Whitehead's claim is, it is less bold than a quite comparable claim made by the eminent historian Sir Herbert Butterfield, referring to that period, the scientific revolution, in which the developments to be recounted in the first two-thirds of this volume form the core. What Butterfield stated was:

> Since [the scientific revolution] overturned the authority in science not only of the middle ages but of the ancient world—since it ended not only in the collapse of scholastic philosophy but in the destruction of Aristotelian physics—it outshines everything since the rise of Christianity and reduces the Renaissance and Reformation to the rank of mere episodes, mere internal displacements, within the system of medieval Christendom. Since it changed the character of man's habitual mental operations even in the conduct of the non-material sciences while transforming the whole diagram of the physical universe and the very texture of human life itself, it looms so large as the real origin both

---

[1] Alfred North Whitehead, *Science and the Modern World* (New American Library, 1956), pp. 46–48.

of the modern world and of the modern mentality that our customary periodisation of European history has become an anachronism and an encumbrance.[2]

Even were one to reject Whitehead's claim that this is humanity's "greatest single intellectual success," it is difficult to deny that this story can boast a cast of main characters that includes some of the most remarkable geniuses who have ever lived: Aristotle, Galileo, Descartes, Newton, and Einstein. Moreover, the materials that follow focus on precisely those achievements that made Galileo, Newton, and Einstein so famous. Although each of these three authors showed genius in many areas, it was what they contributed to mechanics that above all merited their immortality.

Authors of detective or mystery fiction might well envy the historian recounting this tale. Clues are literally spread throughout the universe, from the roll of a child's ball to the revolution of a giant planet, from the flight of a toy balloon to the fall of a giant tree, from the toss of a tennis ball to the trajectory of a massive cannon shell. Numberless as the clues are, omnipresent as they may be, they proved to be so difficult to decipher, so elusive in their significance, that the keenest minds missed their meaning for centuries.

Moreover, the story has an extraordinary denouement. All the various lines of development leading up to the final resolution come together in a manner that is simultaneously exceptionally surprising yet also so remarkably perfect that it commands assent. It is difficult to think of any volume, fictive or factual, that culminates in a more delightful or dramatic climax. Actually, this understates the claim that can be made, for, as we shall see, a double denouement awaits us. First, we shall see Newton reveal a pattern that seems so perfect and so all encompassing that one can scarcely believe that hidden just beneath the surface of the picture painted by Newton lay a still more extraordinary and more convincing configuration, which was eventually brought forth by a man in his mid-twenties, Albert Einstein.

In stating that I believe this is the most remarkable story in secular history, I am aware that others may not find it to be so (especially in my telling of it). One reason they may find it far less satisfying is that comprehending this story requires considerable sophistication, although the ideas are much more accessible than most persons assume. For example, the story can be told in nearly all its richness, even in regard to Einstein's special theory of relativity, without requiring any more mathematics than what one learns in a good high school. But the ideas are sufficiently complex (indeed, they eluded for many centuries nearly all the best minds ever on this planet) that understanding them presents a challenge to everyone except persons coming to

---

[2]Herbert Butterfield, *The Origins of Modern Science 1300–1800,* rev. ed. (Macmillan, 1957), pp. vii–viii.

them with a strong background in mechanics. And that person may face a somewhat different challenge, the challenge of reading earlier authors with the openness of mind necessary to understand how the earlier author saw the world. Moreover, the reader who comes to the texts and ideas presented in this volume with prior experience in mechanics may acquire a deeper understanding of mechanics itself. One reason for this is suggested by Pierre Duhem's remark: "To give the history of a physical principle is at the same time to make a logical analysis of it."[3]

The story to be told is quite instructive; in fact, philosophers of the most diverse sorts have scrutinized it seeking support for their favorite doctrines. One lesson that seems to me to emerge with particular forcefulness from these materials is suggested by a remark made by Pierre Duhem in his classic *Aim and Structure of Physical Theory*.

> ...history shows us that no physical theory has ever been created out of whole cloth. The formation of any physical theory has always proceeded by a series of retouchings which from almost formless first sketches have gradually led the system to more finished states; and in each of these retouchings, the free initiative of the physicist has been counselled, maintained, guided, and sometimes absolutely dictated by the most diverse circumstance, by the opinions of men as well as what the facts teach. A physical theory is not the sudden product of a creation; it is the slow and progressive result of an evolution.
>
> When several taps of the beak break the shell of an egg from which the chick escapes, a child may imagine that this rigid and immobile mass, similar to the white shells he picks up on the edge of a stream, has suddenly taken life and produced the bird who runs away with a chirp, but just where the childish imagination sees a sudden creation, the naturalist recognizes the last stage of a long development; he thinks back to the first fusion of two microscopic nuclei in order to review next the series of divisions, differentiations, and reabsorptions which, cell by cell, have built up the body of the chick.
>
> The ordinary layman judges the birth of physical theories as the child the appearance of a chick.[4]

Duhem was suggesting that an exposure to the history of science not only teaches many new ideas, but also can serve to free persons from erroneous ideas, for example, the notion that major physical theories appear suddenly and with relatively little preparatory work. To illustrate his point, Duhem devoted the next thirty pages of his book to tracing the slow and

---

[3] Pierre Duhem, *The Aim and Structure of Physical Theory,* trans. by Philip P. Wiener (Princeton University Press, 1991), p. 269.
[4] Duhem, *The Aim and Structure,* p. 221.

gradual evolution of the idea of universal gravitation. Another case of im-
mediate relevance for us is the law of inertia, which is simple enough that
Newton stated it in a single sentence. Nonetheless, we shall come to see
that the law of inertia can be compared to a remarkably perfect cathedral
that required the combined labors, extending over centuries, of persons of
extraordinary insight.

The fact that so many hands working over more than twenty centuries
were needed to bring this work to perfection should not obscure the fact that
true genius is involved. Alexander Pope spoke for many when he wrote:
"Nature, and Nature's Laws lay hid in Night:/ God said, *Let Newton be!* and
All was *Light*," [5] We shall see in Newton and in Einstein, and scarcely less
in Galileo, a level of brilliance that is nothing less than awesome. What is
most remarkable, nonetheless, and what perhaps gives the story its greatest interest, is that we shall come to understand their insights, to relive the
process of their discovery, and thereby to see how they as our fellow human
beings attained them.

One special feature of this book is its inclusion of substantial selections
from writings by a number of the most important contributors to the history of mechanics, especially Aristotle, Galileo, Descartes, Huygens, and
Newton. These selections should allow readers to come directly into contact with the thought of these brilliant scientists. Commentaries supplying
explications and contexts accompany these selections, enhancing their ac-
cessibility. Moreover, by reading presentations of these authors in their own
words, readers increase the chance that they will understand not only what
the scientist attained, but also how it was attained. This permits readers to
enter worlds quite different from our own and, as it were, to observe the artist
in the act of creation. Moreover, the included selections provide documentation for the story told in these pages.

Readers may wonder how this book came to be written and what audiences it can serve. The book, which came out of a lifelong interest in the
history of physical science and of mathematics, was originally designed to
serve two groups at the University of Notre Dame: undergraduate students
in Notre Dame's great books program, the Program of Liberal Studies, and
graduate students in Notre Dame's Graduate Program in History and Philosophy of Science. For undergraduates, perhaps half of whom had never
taken a course in physics at any level, it served to introduce them to what
is widely regarded as the most fundamental area in that discipline and also
to the history and philosophy of science. One half of a three credit course

---

[5] Alexander Pope, *Minor Poems,* ed. by Norman Ault, completed by John Butt (Yale University
Press, 1954), p. 317; see also Derek Gjersten, *The Newton Handbook* (Routledge and Kegan
Paul, 1986), p. 439.

proved sufficient time for a careful study of these materials, including the laboratory developed to accompany the Galileo materials. For this group, it was especially important to make the exposition of the physical ideas as clear and accessible as possible. When teaching this text in a graduate course, where the students typically had substantially stronger scientific, historical and philosophical backgrounds, I devoted perhaps one fourth of a course to the text. Teaching in this second context made it especially important that the text be historically correct and based on the best historical research available. Although the differing needs to these two constituencies might seem difficult to serve simultaneously, my experience has been that this can be done and that, moreover, we need more books that are scientifically accessible and also historically sophisticated. The favorable reviews and reasonably brisk sales of two other books[6] that I have written for these same audiences seem to support my view that a need for and interest in such publications exist. In the final revision that I have recently given this book in preparation for its publication, I have made every effort also to serve also the independent reader, young or not so young, who may wish to delve into the lives and thoughts of the great geniuses whose ideas and writings are presented in this volume.

Acknowledgments

I am deeply indebted to five scholars—Dana Densmore, William H. Donahue, André Goddu, Phillip R. Sloan, and Barbara Turpin—for the many contributions they have made concerning the contents of this book. Dana and Bill, both of whom are widely known for their expertise and publications regarding the scientific revolution period, contributed in countless ways to the quality of this volume. Especially helpful suggestions have also been made by Steve Nazaran and by Ryan MacPherson, while serving as teaching assistants for a course based on these materials, and by a student with an especially sharp eye and generous nature, Adam Weaver. Valuable recommendations for improvements were also provided by Keith Lafortune and Brandon Fogel. Cheryl Reed's typing skills contributed significantly to the preparation of the final manuscript for this book. Most recently, Christina Turner, a Notre Dame graduate student, has read the proof copy of the book with great care and insight. Also, Howard J. Fisher of St. John's College (Santa Fe), and Associate Editor at Green Lion Press, has contributed helpful suggestions and careful editing in the final stages of the preparation of this book.

---

[6]M. J. Crowe, *Theories of the World from Antiquity to the Copernican Revolution* (Dover, 1990) and *Modern Theories of the Universe from Herschel to Hubble* (Dover, 1994).

I am indebted to the following persons and/or publishing companies for permission to quote or reprint materials from the sources indicated: Albert Einstein, "Autobiography" in *Albert Einstein: Philosopher-Scientist,* ed. by Paul Arthur Schilpp, vol. 1 (Harper and Row, 1959); Albert Einstein, "On the Electrodynamics of Moving Bodies" in A. Einstein, H. A. Lorentz, H. Weyl, and H. Minkowski, *The Principle of Relativity,* a collection of classic papers trans. by W. Perrett and G. B. Jeffery (Dover republication of the 1923 original); *Aristotle's Physics,* trans. by Hippocrates G. Apostle (Peripatetic Press, 1980); Christiaan Huygens, *On the Motion of Colliding Bodies,* as given in "Christiaan Huygens' *The Motion of Colliding Bodies,*" trans. by Richard J. Blackwell, *Isis, 68* (1977), 574–597; Galileo Galilei, *Dialogue Concerning the Two Chief World Systems—Ptolemaic and Copernican,* trans./ed. Stillman Drake (University of California Press, 1953); René Descartes, *Principles of Philosophy,* trans. by William H. Donahue for this volume from Charles Adam and Paul Tannery (eds.), *Oeuvres de Descartes,* vol. 8.1 (Libraire philosophique J. Vrin, 1973); translations made by William H. Donahue from Isaac Newton's *Mathematical Principles of Natural Philosophy* and included in Dana Densmore, *Newton's* Principia: *The Central Argument,* with translations and illustrations by William H. Donahue (Green Lion Press, 1995); Wall and Emerson, who administer the copyright for Galileo Galilei, *Two New Sciences Including Centers of Gravity & Force of Percussion,* trans. by Stillman Drake (University of Wisconsin Press, 1974).

Michael J. Crowe
Cavanaugh Professor Emeritus
Program of Liberal Studies
University of Notre Dame

Chapter 1

Mechanics before Galileo

Introduction: What Is Mechanics?

Two quite different meanings of the word "mechanics" are encountered.

1. According to the more common of these, mechanics is that area of knowledge concerned with the construction, design, and operation of machines. Thus the person who repairs and maintains automobile or aircraft engines is called a mechanic.

2. The second meaning of the term mechanics is that area of knowledge that treats of motions and tendencies to motions in material bodies. Practitioners of this form of mechanics deal with such topics as how or why falling bodies or thrown baseballs or launched rockets move as they do, or cease moving.

The subject matter treated in this book is mechanics in this second sense.

Mechanics in this second sense is a very abstract and theoretical branch of physics. In the present context, it is important to note that the meaning and extension of the term evolved over time. A good example of this comes from consideration of the definition of mechanics provided in the *American Heritage Dictionary,* where it is described as "The analysis of the action of forces on matter or material systems." The problematic nature of this definition is suggested by the fact that some of the systems of mechanics that we shall be encountering avoid or even deny the legitimacy of the idea of force!

The importance of mechanics in the development of science can scarcely be overestimated. Mechanics is, moreover, frequently described as the most fundamental area of physical science and possibly of all science. Mechanics is not only important for physics and for nearly every area of science, but also is philosophically of exceptional significance. It would be correct to say that throughout history mechanics has been taken to be the paradigm science, the model for all the other areas of science. Philosophers asking the question "what is good science?" or "what is the proper methodology for science?" frequently turn to mechanics. Theological issues, as will become evident in what follows, have also been raised by ideas developed in the history of mechanics.

Some Key Questions Dealt with in Mechanics

Various questions and ideas are central to mechanics. In encountering the history of mechanics, it will prove helpful to have reflected on some of the key questions and ideas in mechanics, including their origins.

Motion: Perhaps the most central idea in mechanics is motion, an idea that seems both simple and clear. We suspect that infants and members of primitive societies have a definite idea of motion, that this concept develops very early in the life of an individual or society. Moreover, if someone asks us early in the morning whether we are moving, we usually reply without hesitation affirmatively or negatively, depending on whether we have left our bed. It turns out, however, that the idea of motion is rather more complex that this implies. Suppose that upon waking up, someone telephones us to find out if we are in motion yet. The answer seems clear enough unless we are on shipboard, and the person is asking whether the ship has yet left port. Then the answer may be that although we have not moved an inch in bed, we are sailing at 30 knots. Or imagine that our caller is resident on the Moon and anxious to find out whether the Earth, the Moon, or the Sun is stationary. If we are Copernicans, then the correct answer is that even without our morning coffee, we are managing to move at over sixty thousand miles per hour. These examples suggest that motion is to some extent a relative term, not so easily defined or measured.

Distance: When do infants attain the idea of distance? Is it when they learn to focus their eyes or to find their hand? How early in life do we attain the idea that some objects are nearer us than others? Related to this issue is the question of how we measure distance. The first stage is no doubt comparative. Children recognize that their parents are taller than they are. All societies eventually develop a standard of distance, possibly the length of the thumb or of the king's foot or the distance to a nearby location. These measures may also be relative: our king has a large foot, your king a small foot. All historians know the complexities of determining what is meant by the standards of length adopted by any given society. Nor have these problems yet been fully resolved, as Americans debate whether to favor feet and miles or meters and kilometers as the standard of distance.

Time: A similar analysis of our notion of time suggests that specifying exactly what constitutes time is a complex matter. It is all too easy living in the twenty-first century in homes with clocks in every room and watches on every (left) wrist to assume that the notion of time is clear and definite. Yet some reflection suggests that experienced time and real time may be quite different; we say "Yesterday just flew by, but today is dragging." Throughout much of the past, humans had only two measures of time: their pulse and the length of the day, the latter failing for those near the north or south poles, where six months separate sunrise and sunset, or on planets rotating and revolving at rates different from the rates for our Earth.

Speed: How early in the life of a child or a society does the notion of speed emerge? Neither Greek ships nor Roman chariots came equipped with speedometers, nor possibly did early Greeks and Romans have a clear notion of the conceptual entity we call speed. A charioteer would be vividly aware that another charioteer moved more rapidly than he did, but would this entail recognition that distance divided by time provides a useful number, which is speed?

Acceleration: Because acceleration is the measure of the change of a speed, it is clearly necessary that persons have the notion speed first. One may wonder what genius first recognized that not only do speeds change, but that one can meaningfully speak of the speed at which a speed changes, i.e., of acceleration. Speed may be observed and possibly even measured without great difficulty, but seeing and measuring accelerations is far more difficult.

It is surely evident that other ideas, e.g., average speed, instantaneous speed, speed of rotation, force, inertia, etc. are important for mechanics. But it is frequently difficult to specify exactly when and how such ideas made their first appearance. Nonetheless, to comprehend the history that follows, to appreciate the achievements of the pioneers of this science, it is important to realize that few if any of these notions come to us as givens; rather they had to be created, improved, and perfected, which processes may take years if not centuries and possibly minds of surpassing brilliance.

Part I: Mechanics in Antiquity[1]

Aristotle

What author in the entire history of physics received the most sustained praise for his ideas? What author received most criticism? A plausible

---

[1]Some of the presentations in this section draw on a discussion of Aristotle's views of mechanics prepared by Professor André Goddu.

answer to the first question is no less plausibly an answer to the second. That answer is Aristotle. Scholars for centuries regarded his physics as among the most profound human creations, but after, say, 1600, they criticized Aristotelian physics with a relentless harshness that was almost without precedent. The paradox evident in this may be relieved somewhat if the physical ideas of Aristotle (384–322 BCE) are approached in the context of the times in which he lived and thought.

When Aristotle was working out the ideas that he presented in his two major treatises on the physical world, his *Physics* and *On the Heavens,* he was very aware of the two main systems of thought that earlier Greek thinkers had formulated concerning physical reality: these were the atomistic and the Platonic systems. Aristotle consciously formulated his ideas in relation to these systems, yet for the most part he advocated a cluster of ideas quite different from those espoused in either system.

Let us look first at the **Atomists**, the most famous of this group being Lucretius (96?–55 BCE), who in his *De rerum natura* expounded ideas and an approach that had been developed centuries earlier. Among the earlier atomists were the pre-Socratic philosophers Leucippus (fl. ca. 480 BCE) and Democritus (d. ca. 361 BCE), whose doctrines were well known to Aristotle. How did atomists view physics? First of all, they refused to draw any distinction between the terrestrial and the celestial regions. Both regions are, they maintained, filled with atoms and the void. Correspondingly, the same methods of inquiry are proper to both areas. In general, they stressed an empirical methodology, rather than a reliance on religious or metaphysical considerations. They also argued that it is not appropriate to ask questions about purpose. In other words, they rejected a teleological approach to nature.

Plato (ca. 427–347 BCE) adopted a very different approach. In common with the Pythagoreans, Plato placed great stress on mathematics. In fact, in his *Timaeus,* Plato urged that the ultimate elements of physical bodies are geometrical. In contrast with the atomists, Plato divided the celestial realm, where the gods dwell, from the terrestrial. Unlike the atomists, he downplayed the importance of sense experience.

The position advocated by **Aristotle** can in some ways be described as intermediary between those two positions. Let us first examine a statement made by Aristotle in his *Physics,* which statement has bearing on his view of the relation of mathematics to physics.

> Having distinguished the various senses of "nature," we should next investigate how the mathematician and the physicist differ with respect to their objects, for physical bodies have also surfaces and solids and lengths and points, and these are the concern of the mathematician. Moreover, is astronomy a distinct science or a part of physics? For

it is absurd that the physicist should understand what the Sun or the Moon is but not what their essential attributes are, not to mention the fact that those who are concerned with nature appear to be discussing the shape of the Moon and of the Sun and to be raising the problem of whether the Earth and the universe are spherical or not.

Now the mathematician, too, is concerned with these, but not insofar as each is a limit of a physical body; nor does he investigate attributes qua existing in such bodies. That is why he separates them, for in thought they are separable from motion; and it makes no difference, nor does any falsity occur in separating them [in thought]. Those who posit Ideas, too, are doing the same but are unaware of it; for they are separating the physical objects [from motion], although these are less separable than the mathematical objects. This becomes clear if one tries to state the definitions in each [science], both of the subjects and of their attributes. For oddness and evenness and straightness and curvature, and also a number and a line and a figure, will each be defined without reference to motion; but not so in the case of flesh and bone and a man, for these are defined like a snub nose and not like a curvature. This is also clear in those parts of mathematics which are more physical, such as optics and harmonics and astronomy, for these are related to geometry in a somewhat converse manner. On the one hand, geometry is concerned with physical lines but not qua physical; on the other, optics is concerned with mathematical lines not qua mathematical but qua physical.

Since we speak of nature in two ways, as form as well as matter, we should investigate the whatness [of the objects of physics] as we would the whatness of snubness. Such objects, then, should be investigated neither without matter nor with respect to matter [alone]. With regard to this we might also raise another problem. Since there are two natures, with which of them should the physicist be concerned? Or should he be concerned with that which has both natures? Of course, if with both natures, then also with each of the two natures. So should the same science be concerned with both natures, or one science with one and another with the other?

If we turn our attention to the ancients, physics would seem to be concerned with matter, for even Empedocles and Democritus touched upon form or essence only slightly. But if art imitates nature and the same science should understand the form and the matter to some extent (for example, the doctor should understand health, and also bile and phlegm in which health exists; the builder should likewise understand the form of the house, and also the matter, namely, bricks and wooden materials; and similarly in each of the other arts), it should be the concern of physics, too, to know both natures.[2]

---

[2] *Aristotle's Physics*, trans. by Hippocrates G. Apostle (Peripatetic Press, 1980), pp. 27–28. See Aristotle, *Physics,* II.2, 193b 23–194a 27.

This passage makes clear that Aristotle rejected the Platonic view that mathematical forms are the constituents of physical bodies. Aristotle believed that Plato had in effect mixed up conceptual categories from two distinct although related disciplines, mathematics and physics. Correspondingly, Aristotle rejected the primacy given to mathematics in the Platonic tradition. Aristotle was also opposed, though for different reasons, to the approach of the atomists. He faults them for giving inadequate attention to the nature or essence of material objects. Aristotle believed it was very important to ask "why" questions about nature. We shall see that some later authors argued that "how" questions are more important, or at least that such questions stand a greater chance of being answered.

The passage from Aristotle also suggests the Aristotelian doctrine that physics and astronomy are distinct disciplines, differing not only in subject matter but also methodology. Elsewhere Aristotle argued that the celestial and terrestrial realms are entirely different. The heavens for Aristotle contain perfect bodies, which move according to laws and patterns quite distinct from those on Earth. For example, the celestial bodies move in circles and come back upon themselves so that they can move eternally. Moreover, the celestial bodies are moved by the divinities associated with them in Aristotle's cosmology. Terrestrial bodies, on the other hand, are imperfect and do not move eternally. All this, of course, fitted very well with Aristotle's belief in the geocentric system of the universe. It is worth noting, nonetheless, that Aristotle's system cannot have been a direct product of his geocentrism. This is clear from the fact that the atomists and Platonists were also convinced of the centrality of the Earth's location.

Aristotle's distinction between the terrestrial and celestial realm extended to the composition of the bodies concerned. His belief was that all terrestrial bodies are composed of earth, air, fire, or water, or a combination of these four elements. These elements, in particular, air and fire, extend up to the region below the Moon but no farther. The celestial realm itself is composed of a perfect and incorruptible material.

Regarding method, Aristotle, again in contrast to the Platonists, opted for a limited empiricism. The task of the physicist is to determine the nature, the essence, of the physical body. When knowledge of this essence is attained, the physicist has the proper knowledge for understanding how bodies change. Although mathematics played a very minor role in Aristotle's physics, it can be suggested that in a sense mathematics entered Aristotelian physics through the back door. Deeply impressed by the power of geometrical thought, Aristotle seems to have formulated his physical system using geometry as a model. This led him to place heavy reliance on deductive thought. In a sense, Aristotle hoped to do for physics what the predecessors of Euclid had done for mathematics. It should be stressed, however, that Aristotle's conception of physics, substantially more than Plato's, assigned a role to experience, to observing carefully how nature behaves.

Aristotle on Place, Motion, and Void

One of the key Aristotelian doctrines is that bodies move toward their **natural place**. Thus Aristotle explained that earthy bodies fall toward the Earth because the natural place for a clump of earth is with the Earth. Similarly, air and fire rise because their natural place lies above us. An important point to note in this is that Aristotle's explanation rests on an analysis of the nature of the body under consideration and what is natural to it.

Concerning motion, which Aristotle viewed as only one form of change, he divided all motions into two categories, natural and violent. When an earthy body falls to the Earth or fire ascends, it is in natural motion, whereas when a heavy projectile is thrown toward the sky, the projectile is in violent motion. In addition, Aristotle maintained that "Every thing in motion is necessarily moved by some thing." [3] For animate bodies, the source of their motion is within the body, whereas for inanimate objects the mover must be external. Aristotle viewed the motion of a body, say a ball thrown in the air, as involving a displacement—the air gives way as the ball passes through it. As Aristotle stated in Book IV of his *Physics,*

> Again, things which are thrown are in motion, though that which pushed them is not touching them, either because of mutual replacement, as some say, or because the air that has been pushed pushes [them] with a motion which is faster than the locomotion of the things pushed, that is, the locomotion with which those things travel to their proper place.[4]

This was quite different from the atomist idea that bodies move into a void. Allied to this is the famous Aristotelian rejection of the atomist belief in the existence of **void** space. Aristotle rejected this atomist doctrine for a number of reasons. One was that he believed that it makes no logical sense to speak of the existence of nothing, which is what, according to the atomists, fills the void spaces in which they believed. But Aristotle had also more empirical arguments against the void. Consider the following analysis.

> We observe that the same weight or body travels faster for two *reasons,* either because there is a difference in the medium through which it travels, as through water or earth or air, or because, other

---

[3] *Aristotle's Physics,* (Apostle translation), p. 127. See Aristotle, *Physics,*, VII.1, 241b 25.
[4] *Aristotle's Physics* (Apostle translation), p. 73. See Aristotle, *Physics,* IV.8, 215a 14–18.

things being the same, the travelling body has an excess of density [or weight] or of lightness. The medium through which the body travels is a cause by the fact that it obstructs that body, most of all if it [the medium] is travelling in the opposite direction, but even if it is *resting*; and it does so more if it is not equally divisible, and such is a more viscous medium.

In this part of his argument against the void, Aristotle is saying that the speed of a body (for example, a ball) moving through some medium (for example, water or air), depends on two factors. One of these is the density of the medium through which the ball moves. It is much harder to throw a ball through water than through air. Aristotle also maintains that the speed of the ball will depend on how heavy the object is; the heavier it is, the more success it will have in pushing the medium aside. He then proceeds to a more quantitative analysis.

The body *A*, then, will travel through medium *B* in time *C* and through medium *D* (which is less viscous) in time *E,* these (*C* and *E*) being proportional to the obstructing medium, if the lengths of *B* and *D* are equal. For let *B* be water and *D* be air. Then the extent to which air is less viscous or less corporeal than water is proportional to the extent to which *A* travels faster through *D* than through *B*. Let the two speeds have the same ratio as that by which air differs from water. Then if air is half as viscous as water, *A* will travel through *B* in twice the time as it will through *D,* and *C* will be twice as long as *E*. And always, the more incorporeal or less obstructive or more easily divisible the medium the faster will the body travel through it.

But there is no ratio in which the void is exceeded by a body, just as there is no ratio of zero (nothingness) to a number. ... [Thus we see that] if, in a given time, a body travels through the least viscous medium a certain distance, the distance it will travel through a void [in the same time] will exceed the other distance by every ratio.[5]

Let us examine what this means. Aristotle is claiming that the speed at which a body moves depends on the medium through which it moves and on the weight of the body. If the distances traversed are the same in two different media (for example, water and air), then the times required to traverse the two media by bodies of the same "weight" will be proportional to the density of the hindering media. Although Aristotle did not do so, this can be expressed by means of the following formula: For two bodies with identical weights ($W_1 = W_2$) traversing the same distances but in different media, the media having densities D_1 and D_2, then the times T_1 and T_2

---

[5] *Aristotle's Physics*, trans. by Hippocrates G. Apostle, pp. 73–74. See Aristotle, *Physics,* IV.8, 215a 25–32, 215b 1–24.

taken by the bodies are related by the formula:

$$\frac{T_1}{T_2} = \frac{D_1}{D_2}.$$

Correspondingly, two bodies of unequal weight will move through one and the same medium with different speeds, those speeds being in direct relation to the differing weights of the two bodies. In other words: If the two bodies move identical distances through media of the same density, then the speeds V_1 and V_2 of those two bodies will satisfy the equation:

$$\frac{V_1}{V_2} = \frac{W_1}{W_2}.$$

We could generalize (again using modern formulas) the conclusions he reached in this selection in the following formula:

$$\frac{V_1}{V_2} = \frac{W_1}{W_2} \cdot \frac{D_2}{D_1}.$$

Aristotle concluded that a vacuum or void is impossible because the speed of the body in a void would be infinite. This follows from the fact that if in the last equation $D_1 = 0$, then V_1 must be infinite. But Aristotle had argued elsewhere that an infinite speed is impossible. Thus Aristotle has provided a quasi-empirical, quasi-mathematical demonstration that it is impossible to have a void space. The flaw in this argument is not easily detected.

In Book VII of his *Physics,* Aristotle made another important statement about violent or forced motion. In particular, he stated:

> ...if A is the mover, B the thing in motion, S the length over which motion has occurred, and T the time taken, (1) in time T a force equal to that of A will cause a thing which is half of B to move over length 2S, and (2) it will cause it to move over length S in half the time of T; for thus there will be a proportion. And (3) if the force of A causes B to move over the length S in time T, it also causes B to move over half of S in half the time of T, and (4) a force equal to half of A causes a thing equal to half of B to move over a length S in time T. For example, let E = A/2 and F = B/2; then the strengths are similarly and proportionately related to the weights, so in equal times they will cause motion over equal lengths.[6]

We can rephrase Aristotle's claims in this passage in the following manner. All other conditions remaining the same,

1. Doubling the force doubles the distance.

---

[6] *Aristotle's Physics* (Apostle trans.), p. 146. See Aristotle, *Physics,* VII.5, 249b 33–250a 9.

2. Doubling the force halves the time.

3. Halving the distance halves the time.

4. Halving the force and also the weight keeps the distance the same.

Aristotle then introduced an important qualification.

> But (5) if E causes F to move over length S in time T, it does not follow
> that in time T it will cause 2F to move over S/2. Thus, if A can cause B
> to move over a distance S in time T, then half of A, which is E, may not
> cause B to move, either for a part of time T over a part of S or over a
> part X of S so related that X : S :: E : A. For it may happen that E will
> not cause B to move at all; for if a given strength causes a quantity of
> motion, half that strength may not cause any quantity of motion, or it
> may not cause a motion in any given time (for if the total strength of
> the shiphaulers is indeed divisible into their number and the distance
> over which they caused the motion, one of them might cause the ship
> to move). It is because of this that Zeno's statement is false when he
> says that any part of the millet causes a sound; for nothing prevents a
> part from failing in any interval of time to move the air which is moved
> when the whole bushel of wheat falls. Indeed, when by itself, it will not
> move even such a part [of the air] as it would when it acts as a part
> of the whole, for in the whole it does not exist except potentially. On
> the other hand, (6) if one force moves one weight and another moves
> another over a given distance S in a given time T, then the forces when
> combined will also move the combined weights over an equal distance
> S in an equal time T, for here a proportion does apply.[7]

Among the points that emerge most clearly from these passages is Aristotle's conviction that violent motion is influenced by the force of the mover. We see also that Aristotle's goal was to provide a causal analysis of motion rather than simply a description of the motion. Moreover, Aristotle's stress on the fact that unless the force attains a certain magnitude, motion will not be produced suggests why he would possibly have resisted expressing his ideas in terms of such formulas as those used a few paragraphs back in explicating his thought.

One of the most important and yet most difficult points to understand about Aristotle is that he advocated physical ideas that are not only significantly different from those now held, but also that his notion of physics differed in important ways from the modern conception of that discipline. A good example of these differences is that the central goal of Aristotle's physical writings was to explain change. When Aristotle discussed such an entity as speed, he was thinking not only of projectiles, but also of the rate

---

[7] *Aristotle's Physics* (Apostle translation), p. 146. See Aristotle, *Physics,* VII.5, 250a 10–28.

of change in the development of a chick embryo. This raises the question of whether it is appropriate to present Aristotle as having a physics and to discuss his system in the same context as Galileo, Newton, and Einstein. The assumption made in this text is that a continuity can be discerned running from the writings of Aristotle on through those of Galileo, Newton, and even Einstein. It is true that under the term *physics,* Aristotle included much that we view as separate from physics and that he neglected or ruled out of consideration substantial parts of what we designate as physics. On the other hand, it is clear from the passages cited that Aristotle had attained a sophisticated understanding of such ideas as time, distance, and speed.

Brief as this analysis of Aristotle's physical ideas has been, it is perhaps sufficient to show that two widespread characterizations of Aristotle's physics are misleading. These claims are that Aristotle sometimes neglected an empirical approach and that he avoided a mathematical and/or quantitative approach. Regarding the first claim, we have seen that Aristotle did observe the properties of bodies; one might almost argue that his approach was excessively empirical. Regarding the second claim, Aristotle's analysis of motion makes evident that he did at times adopt a quantitative and mathematical approach.

Overall, Aristotle in his physical writings provided later authors with an insightful analysis of the motion of terrestrial bodies. Later Greek authors recognized the brilliance of his presentations as did many medieval authors. In fact, Aristotle's physical system was widely recognized for nearly two millennia as the most impressive physical system ever created.

Mechanics in Later Antiquity

One should not assume that the above discussion exhausts the contributions made to mechanics in antiquity. It is well worth noting that Aristotle discussed other issues in mechanics and that various later authors such as Archimedes of Syracuse and Hero of Alexandria made important studies of diverse aspects of mechanics. Archimedes, for example, made very impressive analyses of how to determine specific gravity, of various problems in hydrostatics, and of the laws governing the mechanical device known as the lever.

Aristotle's ideas did not go unchallenged. The most impressive critique of his views made in late antiquity was that of John Philoponus (ca. 500 CE), a convert to Christianity, who challenged, for example, Aristotle's claim that the continued motion of projectiles after they have left the hand of the thrower is to be explained by the action of the air pressing on the object. Even more importantly, Philoponus questioned Aristotle's claim that the rate at which bodies fall is proportional to their weight. Philoponus asserted in reference to that claim:

> But this is completely erroneous, and our view may be corroborated by actual observation more effectively than by any sort of verbal argument. For if you let fall from the same height two weights, of which one is many times as heavy as the other, you will see that the ratio of the time required for the motion does not depend on the ratio of the weights, but that the difference in time is a very small one.[8]

This passage suggests how different the actual history of mechanics is from the history attributed to it in popular thought, for this passage makes clear that the widespread belief that Galileo was the first to propose testing Aristotle's claim by dropping bodies of differing weights from some high place, e.g., the tower in Pisa, is off by more than a millennium. It is also significant that there is no sound evidence that Galileo ever performed the experiment of dropping weights from the leaning tower of Pisa.

Part II: Mechanics in the Medieval Period

The Mertonians, Oresme, and the Mean Speed Theorem

What level of sophistication did science attain in the medieval period? It is widely believed that science sank to a very low level during the middle ages, a level from which it recovered only with the scientific revolution of the sixteenth and seventeenth centuries, when such scientists as Galileo formulated new theories. One exemplification of this view appears in Henry Smith Williams's *The Great Astronomers*. Chapter VI of that volume is entitled: "THE CHRISTIAN WORLD—TWELVE CENTURIES OF PROGRESS (325–1543, A.D.)." Williams opened the chapter by stating:

> From the Council of Nicea, at which the Emperor Constantine made Europe safe for Athanasian Theocracy, to the time of Copernicus, whose great work, teaching that earth is not the centre of the universe, was to remain under ban of the Council of the Inquisition until fifteen centuries after the Nicene victory, the record of astronomical progress in all Christendom may most charitably be expressed in these words.[9]

That Williams intended that the title he gave this chapter be read as sarcasm is evident from the fact that he left the next four pages blank, or not quite blank. At the bottom of the second page Williams inserted a note: "The reader will note that this page, as well as the next two, are blank. This is

---

[8] As given in Morris R. Cohen and I. E. Drabkin (eds.), *A Source Book in Greek Science* (Harvard University Press, 1958), p. 220.
[9] Henry Smith Williams, *The Great Astronomers* (Newton, 1932), p. 99.

the most graphic method the author can devise to show that, astronomically speaking, nothing happened during the Middle Ages."[10]

Historians of science have long since recognized that Williams's characterization of the medieval period is seriously wrong. In fact, Williams himself should have recognized this because years before the publication of Williams's volume, the first five volumes of Pierre Duhem's ten-volume *Le système du monde: Histoire des doctrines cosmologiques de Platon à Copernic* (*The System of the World: History of Cosmological Doctrines from Plato to Copernicus*) (Hermann, 1913–1959) had shown the incorrectness of such a view of the middle ages. One bit of evidence of the defectiveness of Williams's position about medieval science comes from examining a development in mechanics that occurred in the fourteenth century. What is striking about this case is that scholars have shown that one of the keystones of mechanics that had long been credited to Galileo had been discovered nearly three centuries earlier. Moreover, these medieval authors had by then formulated some of the fundamental definitions and mathematical relations of mechanics. Let us examine what these authors achieved.

A group of medieval scholastics located at Merton College of Oxford University and at the University of Paris by about 1350 not only had thought deeply about the problems of motion, but also had succeeded in attaining some ideas that for many years were listed among discoveries attributed to Galileo. They had, in particular, discovered a crucially important relationship between acceleration, average speed, and the distance traversed by a moving body.

We shall examine this discovery, but only in sufficient detail to suggest the legitimacy of the claim that these medieval scholastics attained a number of the definitions and mathematical relations fundamental to mechanics. Let us consider a passage from a work written around 1350 by **Nicole Oresme** (ca. 1325–1382), who taught at the University of Paris and eventually became Bishop of Lisieux. In reading this passage from Oresme, keep in mind that these medieval scholastics were deeply involved with the broad question of the "**intension and remission of forms**," that is, they sought a precise, quantitative method for discussing how forms or qualities increase and decrease. Typical of the forms they discussed are the redness of ripening apples, the charity in a person's soul, the heat in a container of water over a fire, and the speed of a body. They realized that such a form or quality could:

1. remain constant,

2. change at a constant rate, or

---

[10] Williams, *Astronomers*, p. 100.

3. change at a nonconstant rate.

They described the second type of change as "uniformly difform," the third as "difformly difform." If applied to motion, #2 would in our terms be a constant rate of change and #3 a nonconstant rate of change. Moreover, they represented such changes of form graphically.

These features appear in the following passage from Oresme's *Treatise on the Configurations of Qualities and Motions*. Furthermore, this passage contains what one historian of medieval science has described as "probably the most outstanding single medieval contribution to the history of physics."[11] This is the so-called **Merton Mean Speed Theorem**. Expressed in somewhat modern terms, the chief question that Oresme wished to answer was: For a uniformly accelerating body, how are average speed and distance traversed related to each other? Oresme stated:

> Every quality, if it is uniformly difform, is of the same quantity as would be the quality of the same or equal subject that is uniform according to the degree of the middle point of the same subject. I understand this to hold if the quality is linear. If it is a surface quality, [then its quantity is equal to that of a quality of the same subject which is uniform] according to the degree of the middle line; if corporeal, according to the degree of the middle surface, always understanding [these concepts] in a conformable way. This will be demonstrated first for linear quality. Hence let there be a quality imaginable by $\triangle ABC$ [see diagram], the quality being uniformly difform and terminated at no degree in point B.

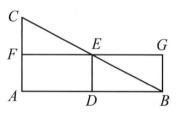

> And let D be the middle point of the subject line. The degree of this point, or its intensity, is imagined by line DE. Therefore, the quality which would be uniform throughout the whole subject at degree DE is imaginable by rectangle $AFGB$, as is evident by the tenth chapter of the first part. Therefore, it is evident by the 26th [proposition] of [Book] I [of the *Elements*] of Euclid that the two small triangles EFC and EGB are equal. Therefore, the larger $\triangle BAC$, which designates the uniformly difform quality, and the rectangle $AFGB$, which designates the quality uniform in the degree of the middle point, are equal. Therefore the qualities imaginable by a triangle and a rectangle of this kind are equal. And this is what has been proposed.[12]

[11] Edward Grant, *Physical Science in the Middle Ages* (John Wiley, 1971), p. 56.

[12] Nicole Oresme, "Treatise on the Configurations of Qualities and Motions" as given in Marshall Clagett (ed. and trans.), *Nicole Oresme and the Medieval Geometry of Qualities and Motions* (University of Wisconsin Press, 1968), pp. 409–411.

Commentary

What Oresme did in this passage, if translated into modern terms, is to develop an important equation that applies to uniformly accelerated motions; in particular, he showed that for s = distance, v_f = final velocity, t = time, and v_{ave} = average velocity, we have:

$$s = \frac{v_f}{2} \cdot t = v_{ave} \cdot t.$$

Expressed in verbal form, he showed that the same distance is traversed in two very different cases: first, the case where the speed of a body increases from an initial speed $v_o = 0$ to a final speed v_f, and second, the case where the speed remains constant, in particular, where it has the value $\frac{v_f}{2}$. To see this, note that in the rectangle $ABGF$, DE represents a constant speed, specifically, the average speed, whereas AB represents time. Similarly, in $\triangle ABC$, AB again represents time, whereas the line AC represents the final speed of the body, which has increased at a constant rate from 0, the 0 value being represented by the point B. Oresme claims that because the area of rectangle $ABGF$ is equal to the area of $\triangle ABC$, the distances resulting from these very different speeds (both acting over the same period of time AB) are themselves equal. Phrased in a more formal manner, the argument proceeds from the facts that area $\triangle ABC = \frac{AB \cdot AC}{2}$ and that the area of rectangle $ABGF = AB \cdot ED$. Then we make the following substitutions: $v_f = AC$, $t = AB$, and $v_{ave} = ED$ to get the formula that (for constant acceleration):

$$s \text{ (distance)} = \frac{v_f \cdot t}{2} = v_{ave} \cdot t.$$

Oresme went on to state:

In the same way it can be argued for a quality uniformly difform terminated in both extremes at a certain degree [see diagram], as would be the quality imaginable by quadrangle $ABCD$.

For let line DE be drawn parallel to the subject base and $\triangle CED$ would be formed. Then let line FG be drawn through the degree of the middle point which is equal and parallel to the subject base. Also, let line GD be drawn. Then, as before, it will be proved that $\triangle CED =$ 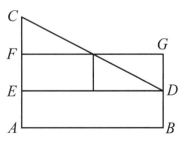 rectangle $EFGD$. Therefore, with the common rectangle $AEDB$ added to both of them, the two total areas are equal, namely quadrangle $ACDB$, which designates the uniformly difform quality, and the rectangle $AFGB$, which would designate the quality uniform at the degree of the middle point of the subject AB. Therefore, by chapter ten of the first part, the qualities designatable by quadrangles of this kind are equal.[13]

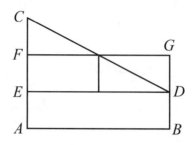

Commentary

In this passage, Oresme has in effect proven the now well known formula

$$s = \frac{v_o + v_f}{2} \cdot t = v_{\text{ave}} \cdot t.$$

As is evident from his statements, $AB = ED = t$, $v_f = AC$, $v_o = DB$, and $BG = v_{\text{ave}}$. These substitutions are then made in his first equation to achieve his result. In particular, knowing that

area quadrangle $ABDC$ = area rectangle $ABGF$,

we can write

area quadrangle $ABDC$ = area rectangle $ABDE$+ area $\triangle EDC$
= area rectangle $ABGF$.

From this we see that

$$AB \cdot AE + \frac{ED \cdot EC}{2} = AB \cdot AF.$$

By making the appropriate substitutions, this becomes

$$t \cdot v_o + \frac{t \cdot (v_f - v_o)}{2} = t \cdot \left(\frac{v_f}{2} + \frac{v_o}{2}\right) = t \cdot v_{\text{ave}},$$

or

$$\frac{v_f + v_o}{2} \cdot t = v_{\text{ave}} \cdot t,$$

which is what we wished to prove.

---

[13]Oresme, "Configurations," p. 411.

It is important to note a point that is implicit in Oresme's analysis and that he made explicit in a subsequent discussion. This is that the area enclosed by these figures in which one side represents time and the other side speed, is numerically equal to the distance covered. Furthermore, Oresme realized that this holds true even if the plot of speed versus time is curved, as is the case for a difformly difform change of motion, that is, for a non-constant acceleration. Mathematical complexities, however, make it difficult to determine the area under a curve. Later, much work was done on this problem, work that played an important role in the creation of calculus.

It is also worth stressing that Oresme attained a notion quite near to what in modern physics is meant by acceleration and also that Oresme's results are mathematical in form. It remains to be investigated whether any motions that are observed in nature obey these formulas. This must be so if cases occur in nature in which there is a constant acceleration. Oresme and the Mertonians did not actively pursue this question. Part of the reason that they did not pursue it was that the goal of their analysis was not primarily to understand the motion of bodies, but rather their concern was far more general: to understand change itself, whether it be the motion of bodies or the growth in grace in a person's soul!

One further point can be drawn from this discussion. As is well known, for a constant acceleration a acting over a time t on a body with an initial speed v_0, we have that $v_f = v_0 + at$. If we substitute this formula for v_f into the second of the two equations derived by Oresme, we see that for a constant acceleration a:

$$s = \frac{v_o + v_f}{2} \cdot t = \frac{v_0 + (v_0 + at)}{2} \cdot t = v_0 \cdot t + \frac{at^2}{2}.$$

For cases where $v_o = 0$,

$$s = \frac{at^2}{2}.$$

We shall come to see that this is a very useful equation.

Sample Problems

1. Suppose an airplane constantly accelerates from 0 mph to 600 mph over a period of 4 minutes. Calculate the distance covered.

Solution: In solving this, we can use either of two equations:

1. $s = \dfrac{v_o + v_f}{2} \cdot t$

or

2. $s = \dfrac{at^2}{2}$ or $s = \dfrac{1}{2} \cdot a \cdot t^2.$

First case:

$$s = \frac{600 \text{ mph} + 0 \text{ mph}}{2} \cdot 4 \text{ min.} = 300 \text{ mph} \cdot \frac{1}{15} \text{ hr.} = 20 \text{ miles.}$$

Second Case:

$$\text{The acceleration is equal to} \quad \frac{600 \text{ mph}}{4 \text{ min.}} = \frac{150 \text{ mph}}{\text{min.}}$$

$$
\begin{aligned}
\text{Hence} \quad s &= \frac{1}{2} \cdot \frac{150 \text{ mph}}{\text{min.}} \cdot (4 \text{ min.})^2 \\
&= \frac{75 \text{ mph}}{\text{min.}} \cdot 16 \text{ min.}^2 \\
&= 1{,}200 \cdot \frac{1}{60 \text{ min.} \cdot \text{min.}} \cdot \text{miles} \cdot \text{min.}^2 \\
&= 20 \text{ miles.}
\end{aligned}
$$

2. Suppose a rocket accelerates from rest at the rate of 10 mph per second for 10 minutes. How far will it have gone at the end of 10 minutes?

Solution:

$$
\begin{aligned}
s &= \frac{at^2}{2} \text{ or } s = \frac{1}{2} \cdot a \cdot t^2 \\
&= \frac{1}{2} \cdot \frac{10 \text{ mph}}{\text{sec.}} \cdot (10 \text{ min.})^2 \\
&= \frac{1}{2} \cdot \frac{10 \text{ mph}}{\text{hr.} / 3600} \cdot \frac{1}{36} \text{ hrs.}^2 \\
&= 5 \cdot 100 \text{ miles} = 500 \text{ miles.}
\end{aligned}
$$

3. Suppose a train at some instant is traveling at 30 mph. It accelerates at the rate of 5 mph per minute for 6 minutes. How far will it travel during those six minutes?

Solution: This corresponds to Oresme's second case. The formula is

$$s = v_o t + \frac{at^2}{2}$$

$$= 30 \text{ mph} \cdot \frac{1}{10} \text{ hr.} + \frac{1}{2} \cdot \frac{5 \text{ mph}}{\text{min.}} \cdot (6 \text{ min.})^2$$

$$= 3 \text{ miles} + \frac{1}{2} \cdot \frac{5 \text{ miles}}{60 \text{ min.} \cdot \text{ min.}} \cdot 36 \text{ min.}^2$$

$$= 3 \text{ miles} + \frac{300}{200} \text{ miles}$$

$$= 4.5 \text{ miles.}$$

4. Derive a formula relating initial speed, final speed, distance, and acceleration.

Solution: We know

$$a = \frac{v_f - v_o}{t} \quad \text{and} \quad s = \frac{v_f + v_o}{2} \cdot t.$$

We shall solve this by eliminating t from these two equations.
From the first equation, we have

$$t = \frac{v_f - v_o}{a}.$$

Substituting this into the second equation gives

$$s = \frac{v_f + v_o}{2} \cdot \frac{v_f - v_o}{a}.$$

Hence

$$2as = v_f^2 - v_o^2.$$

This equation is useful in solving various problems.

Summary

We are now in possession of a number of useful equations, which can aid in analyzing motion. These are:

$$1) \quad v = \frac{s}{t}$$

$$2) \quad v_{ave} = \frac{v_f + v_o}{2}$$

$$3) \quad a = \frac{\Delta v}{\Delta t}$$

$$4) \quad v_f = v_o + at$$

$$5) \quad s = v_o t + \frac{at^2}{2}$$

$$6) \quad 2as = v_f^2 - v_o^2$$

Moreover, we have seen that these results, except for #6 and to some extent #5, were known to a number of late medieval scholastics. The recognition of this fact, which was attained only in the twentieth century, has significantly altered historians' conceptions of the middle ages.

An important question to ask about these equations concerns the source from which each comes or, to put it differently, its epistemological status. One or more of the equations may in effect be a definition; another possibility is that it could be an empirical result, which would need to be checked in a laboratory. Or the equation could be a mathematical derivation from one or more other equations and definitions. A careful analysis of the first equation reveals that rather than being a law of nature, it is a definition of what we mean by the term *speed*.

It is important to realize the limitations on this analysis of motion. First of all, it is **kinematic**, not dynamic; in other words, it can be used only to describe motion, but not to explain it. Second, although it appears to be broadly useful, it would be necessary, were we to wish to employ it, to show that it is, in fact, applicable to nature. For example, it is certainly not obvious that it can be directly applied to bodies in free fall or that it can be used to analyze projectile motion. Consequently, we must as yet investigate **whether** and **how** it can be applied in such cases.

Part III:
Two Major Problems in Early Modern Mechanics

Anyone in early modern times wishing to work out a science of mechanics had to contend with two very serious and fundamental questions. These not unrelated questions were: (1) does the Earth itself move? and (2) is motion purely relative? To see the severity of the first question, one can ponder the problems one would encounter in trying to study the motions of a set of billiard balls placed on a moving carousel. If the Earth rotates on its axis, revolves around the Sun, or has other motions, these motions would certainly complicate the task of determining how objects located on the Earth move. The problem of the relativity of motion is hardly less severe. It would seem that unless one can locate a fixed reference point, one cannot tell even whether a given body is in motion. Galileo and his contemporaries interested in mechanics (as well as many of their predecessors) had to deal with these problems.

The Problem of the Possibility of the Earth's Motion

Already in Greek antiquity, various authors had raised the question of whether the Earth may rotate on its axis. Such authors realized that many celestial appearances would be the same if the starry vault rotates and the Earth is motionless at its center, or if the Earth rotates and the starry vault remains fixed in position. Moreover, in the third century BCE, Aristarchus of Samos suggested that the Earth not only rotates but also that it revolves around the Sun. Numerous objections could be made to these bold claims, not least that such motions would disrupt the motion of bodies on the Earth.

Although the heliocentric hypothesis of Aristarchus found few if any supporters in antiquity, it came to life again in the sixteenth century, when, in 1543, Nicholas Copernicus published his *De revolutionibus orbium coelestium*. In that work, the famous Polish astronomer laid out an array of arguments in support of this startling hypothesis. Relatively few sixteenth-century intellectuals adopted the Copernican system; in fact, it has been estimated that even by the end of the sixteenth century, only ten persons had converted to the Copernican system.[14] One reason why so few favored the Copernican theory was that it seemed to contradict known laws and principles of mechanics. In fact, if it were established practice that one could gain recognition in an area of science just for raising major difficult questions, even if the person were unable to find answers for most of those questions, then Nicholas Copernicus would deserve credit for having made an extremely valuable contribution to mechanics. Contemporaries of Copernicus realized that if they were to adopt his astronomical system, they would have to abandon much of traditional physics. Let us consider a few of these difficulties.

Using data on the size of the Earth available at the time of Copernicus, one could conclude that if the Earth rotates every twenty-four hours, then a

---

[14] Robert Westman, "Astronomer's Role in the Sixteenth Century: A Preliminary Study," *History of Science, 17* (1980), 105–147:136, n. 6.

person on the equator must be moving at about 1,000 mph. Moreover, if one factors in the speed of the Earth due to its revolution around the Sun, then one gets a far higher velocity. If one makes the calculation using modern values for the Earth-Sun distance, one finds that we are moving at about 67,000 mph. Copernicus's value for the Earth's distance from the Sun was about twenty times too low. Nonetheless, using his figures, the speed given the Earth by its revolution around the Sun is about 3,200 mph. Why are not these huge speeds evident? Why are not clouds and birds left behind as the Earth rotates and revolves? Why do cannon balls travel the same distance whether they are shot in the direction of the Earth's motion or against it? These were very serious problems for sixteenth- and early seventeenth-century scientists, at least for those who in any way favored the Copernican hypothesis.

Another serious difficulty relates to the fact that the Copernican theory, by turning the Sun into a star, deprived the Earth of its central position in the universe, and thus the Aristotelian idea of bodies moving toward the center of the world fell to pieces. More generally, the Aristotelian distinction between the terrestrial realm and the celestial realm was in effect called into question by the Copernican theory.

The Problem of the Relativity of Motion

Many Greek philosophers, especially Plato and Aristotle, were troubled by the problem of relativism. As early as the sixth century BCE, Xenophanes of Colophon had indicated the relativism of Greek religious notions by remarking: "Aethiopians have gods with snub noses and black hair, Thracians have gods with grey eyes and red hair."[15] Democritus, writing a century later, also pointed to the relativity of tastes: "Sweet exist by convention, bitter by convention, colour by convention, and atoms and Void [alone] exist in reality."[16] Protagoras of Abdera, also late fifth century BCE, made the claim even more strongly: "Of all things the measure is man. . . ."[17]

One particular form of relativism that has been especially significant for science is relativity of motion. Is motion purely relative, or is there a way of defining the motion of a body that is independent of the objects around it? Sitting in a train station and looking out, we see the train next to us in motion—but then, seeing the platform pass us by, we recognize that *we* are in motion. The problem of relativity of motion also applies to speeds. If

---

[15] As quoted in Kathleen Freeman (ed.), *Ancilla to the Pre-Socratic Philosophers* (Harvard University Press, 1956), p. 22.

[16] Freeman, *Ancilla,* p. 93.

[17] Freeman, *Ancilla,* p. 125.

we find that we are moving at seventy miles an hour, we are comfortable if in a car, train, or bus, but if we are jumping out of a tree, riding a bike, or rocketing to the Moon, reports of a speed of seventy miles an hour cause alarm. It is perhaps impossible to tell what author first discussed the relativity of motion; in fact, the history of ideas of the relativity of motion does not seem to have attracted extensive attention from historians.[18] But it is difficult to imagine that the fifth century BCE philosopher Zeno of Elea, famous for his paradoxes of motion, was not aware of the problems associated with the relativity of motion.

Various passages from Aristotle's writings show that he had thought deeply about problems of relativity of motion. For example, he commented in his *Physics:*

> Moreover, the locomotion of physical bodies and of simple bodies (e.g., of fire, earth, and the like) make it clear not only that a place is some thing, but also that it has some power. For each of those bodies, if not prevented, travels to its own place, some of them up and others down, and these (up and down and the rest of the six directions) are parts and species of place. Now such directions (up, down, right, left, etc.) do not exist only relative to us; for to us a thing is not always the same in direction but changes accordingly as we change our position, whichever way we may happen to turn, and so the same thing often is now to the right, now to the left, now up, now down, now ahead, now behind. By nature, on the other hand, each of these is distinct and exists apart from the others; for the up-direction is not any chance direction but where fire or a light object travels, and likewise the down-direction is not any chance direction but where heavy or earthy bodies are carried, as if these directions differed not only in position but also in power.[19]

This passage suggests that earlier authors had raised the question of the relativity of motion and that Aristotle wrote this passage to argue that motion is not relative. This passage also seems to be based on Aristotle's cosmology—in particular, on his notion that the Earth is at the center of the universe, which extends up to the spherical vault of the stars, which is the outermost limit of this large but finite universe.

That the idea of relativity of motion was widely recognized among ancient authors is indicated by a passage from Virgil's *Aeneid,* in which Aeneas

---

[18]Two exceptions are Julian Barbour, *Absolute or Relative Motion? A Study from a Machian Point of View of the Discovery and the Structure of Dynamical Theories,* vol. 1: *The Discovery of Dynamics* (Cambridge University Press, 1989) and Banesh Hoffmann, "Relativity" in *Dictionary of the History of Ideas,* ed. by Philip Wiener, vol. 4 (Charles Scribner's Sons, 1973), pp. 74–92.

[19]*Aristotle's Physics,* trans. by Hippocrates G. Apostle (Peripatetic Press, 1980), pp. 59–60. See Aristotle's *Physics,* IV.2, 208b 9–23.

states: "Forth from the harbor we sail, and the cities slip backwards."[20] The problem of the relativity of motion was perhaps felt most intensely in antiquity by those astronomers who recognized that the motions of the starry vault could be explained either by having the Earth rotate while the starry vault remains fixed, or vice versa. The most famous ancient astronomer Claudius Ptolemy in his *Almagest,* written in the second century CE, showed his awareness of this problem while discussing the question whether it is possible that the Earth could move.

> But certain people, [propounding] what they consider a more persuasive view, agree with the above, since they have no argument to bring against it, but think that there could be no evidence to oppose their view if, for instance, they supposed the heavens to remain motionless, and the earth to revolve from west to east about the same axis [as the heavens], making approximately one revolution each day; or if they made both heaven and earth move by any amount whatever, provided, as we said, it is about the same axis, and in such a way as to preserve the overtaking of one by the other. However, they do not realise that, although there is perhaps nothing in the celestial phenomena which would count against that hypothesis, at least from simpler considerations, nevertheless from what would occur here on earth and in the air, one can see that such a notion is quite ridiculous.[21]

That he took the issue seriously is evident from the number of arguments he presented in this quotation and elsewhere in his *Almagest* to prove that the Earth does not actually move.

Although the question of the relativity of motion, especially in the context of the possibility of the Earth's motion, was discussed by a number of medieval intellectuals, including Thomas Aquinas,[22] possibly the two most thorough discussions penned in the middle ages of the problem of the relativity of motion were those by Jean Buridan (ca. 1300–ca. 1358) and Nicole Oresme (ca. 1325–1382).

Buridan, in his *Questiones super libris quattuor de caelo et mundo,* at one point treated the question "whether the earth is always at rest in the center of the universe." In dealing with this issue he remarked that

---

[20] It is certainly interesting and significant that this passage was cited by Copernicus; see Nicholas Copernicus, *On the Revolutions,* edited by Jerzy Dobrzycki and translated with commentary by Edward Rosen (Johns Hopkins University Press, 1978), p. 16. See Virgil's *Aeneid,* III, 72.

[21] As quoted from Ptolemy's *Almagest* (Bk. 1, Ch. 7) in M. J. Crowe, *Theories of the World from Antiquity to the Copernican Revolution,* 2nd ed. (Dover, 2001), p. 62.

[22] For a relevant passage from Aquinas's commentary on Aristotle's *De Caelo,* see Edward Grant (ed.), *Source Book in Medieval Science* (Harvard University Press, 1974), pp. 499–500.

> many people have held as probable that it is not contradictory to appearances for the earth to be moved circularly... and that on any given natural day it makes a complete rotation from west to east by returning again to the west. Then it is necessary to posit that the stellar sphere would be at rest, and then night and day would take place through such a motion of the earth, so that that motion of the earth would be a diurnal motion (*motus diurnus*). The following is an example of this [kind of thing]. If anyone is moved in a ship and he imagines that he is at rest, then, should he see another ship which is truly at rest, it will appear to him that the other ship is moved. This is so because his eye would be completely in the same relationship to the other ship regardless of whether his own ship is at rest and the other moved, or the contrary situation prevailed.[23]

After discussing a number of arguments concerning whether the Earth does rotate, Buridan concluded that the evidence points toward the traditional position of the Earth's immobility.

The discussion of this issue by Nicole Oresme in his *Le livre du ciel et du monde* is even more impressive. In fact, he went so far as to state regarding the idea of the Earth's rotation that "it is impossible to demonstrate from any experience that the contrary is true."[24] Part of his argument for this claim consists of an analysis of what one would see if located on a ship moving smoothly alongside another ship with identical motion. In this situation, what one sees of the second ship will lead one to the conclusion that both ships are stationary. Moreover, Oresme maintains that if today the Earth were to be stationary and the starry vault were to rotate as traditionally assumed and tomorrow the reverse occurred, "we should not be able to perceive this change, and everything would appear exactly the same both today and tomorrow with respect to this mutation."[25] Oresme's discussion becomes even more impressive when he argues against the claim that if the Earth were to rotate, then an arrow shot up perpendicularly to the Earth's surface should fall some distance away. In countering this objection, he returned to motion on shipboard, suggesting than an arrow shot directly upward from a moving ship will partake of the ship's motion and consequently land at the point from which it was shot. Like Buridan, Oresme ultimately sided with the traditional view that the Earth does not move, citing scripture in support of this conclusion. As we shall see later, arguments very similar to these were fundamental to Galileo's epoch-making contributions to mechanics.

In concluding this discussion of the relativity of motion, two comments

---

[23] As given in Grant, *Source Book,* pp. 500–501. For an interesting discussion of this text, see Barbour, *Absolute or Relative Motion?* pp. 203–206.

[24] As given in Grant, *Source Book,* p. 504.

[25] As given in Grant, *Source Book,* p. 505.

should prove useful in linking it with later discussions. First, although all these authors showed an understanding of the problem of relative motion, none opted for the view that motion is purely relative, that there is no way to decide whether any given body is stationary or in motion. Second, consideration of these authors suggests a thesis worth exploring in the materials that follow. The thesis concerns whether and how considerations of the relativity of motion affected the development of mechanics. On the one hand, it could be claimed that this difficult problem retarded the advancement of mechanics. It is surely true that it complicated many analyses of motion and confused not a few intellectuals as they pondered the nature of motion. The thesis suggested here is very different. It is that recognition of the relativity of motion and of its ramifications was a major source of advancements in mechanics, that a number of the main advances that occurred in mechanics were directly linked to an enriched understanding of the relativity of motion. An especially striking instance of this, discussed in subsequent chapters, is the law of inertia.

Some Problems to Ponder

1) If there were only one body in the universe, would it be possible to move it?

2) If there are only two bodies in the universe and we know that only one is movable, would it be possible to determine which of the two bodies it is that can be moved?

3) Suppose that there exist only two bodies in the universe: one is an immense sphere, the interior surface of which is decorated with hundreds of small points of light, and the other is a very small spherical planet located at the center of this starry vault. Let us assume that the sphere rotates once every 24 hours, whereas the planet remains fixed in position, i.e., it neither rotates nor revolves. What motion would an inhabitant of the starry sphere, convinced that the starry vault is at rest, attribute to the planet at the center? Represent this by means of a diagram. Are there any conclusive arguments that an inhabitant of the planet could formulate to prove that the sphere is rotating? Or are there arguments that the inhabitant of the starry sphere could use to show that the planet is rotating?

4) Suppose only two bodies, A and B, exist in the universe. Suppose that the inhabitants of B see A move around B in a circle in one year. Let us suppose that inhabitants of A believe that A is motionless. Specify the period and the shape of the orbit that they will attribute to B. Let us assume that the body A

always retains the same orientation; that is, were there actually a letter "A" on top of it, that letter would continue to remain right side up. Put another way, assume that body A does not rotate. Are there any arguments that inhabitants of B could use to prove to inhabitants of A that it is their body that is actually revolving?

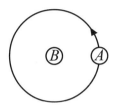

Galileo Galilei

Chapter 2

Galileo and Terrestrial Mechanics

Chronology of Galileo's Life

1564   Birth of Galileo Galilei in Pisa on February 15, the son of a cloth merchant and musician, Vincenzio Galilei.

1575   Begins a period of study lasting until 1581 at the monastery of Santa Maria at Vallombrosa; for a time becomes a novice in the Vallombrosan order.

1581   Enrolls at the University of Pisa, initially for medical studies, but soon turns to mathematics.

1583–8 During this period, Galileo discovers the isochronism of the pendulum, i.e., that the time of swing of a pendulum of a particular length is independent of the amplitude of the swing.

1589   Having by this time composed a treatise on the center of gravity in bodies and invented a hydrostatic balance, he becomes mathematics professor at Pisa.

1592   Becomes professor of mathematics at the University of Padua.

1597   In a letter to Kepler, Galileo states that he had accepted Copernicanism a few years earlier.

1609   Hearing of a device constructed by a Flemish optician (Hans Lipperhey) by which distant objects can be seen as if nearby, Galileo makes his own telescope and begins using it for astronomical observation.

1610   Publishes his *Sidereus Nuncius* (*Starry Messenger*) in which he announces his discovery of the mountainous surface of the Moon, the four moons of Jupiter, and the vast number of stars seen by means of the telescope. Resigns from Padua University;

	moves to Florence with Grand Duke Cosimo II de Medici as patron.
1613	Publishes *Letters on Sunspots* in which he announces his discovery of the phases of Venus, the composite nature of Saturn, and sunspots. By this time, his Copernicanism is beginning to cause controversy.
1615	Cardinal Robert Bellarmine writes to Galileo, urging him to present Copernicanism only in a hypothetical manner.
1616	Holy Office bans Copernicus's book until certain passages are corrected; Galileo warned not to defend Copernicanism.
1620	Congregation of the Index specifies corrections that, if observed, allow reading of Copernicus's book.
1621	Pope Paul V, Cardinal Bellarmine, and Cosimo de Medici all die.
1624	Travels to Rome in hopes of persuading the new pope, Urban VIII, to allow him to write on the Copernican system. Pope agrees, provided Galileo treats Copernicanism hypothetically.
1632	Publishes his *Dialogo. . .sopra i due massimi sistemi del mondo: Tolemaico, e Copernicano* (*Dialogue concerning the Two Chief World Systems—Ptolemaic and Copernican*), vigorously advocating the Copernican system.
1633	Trial of Galileo in which his *Dialogue* is banned; Galileo abjures Copernican system.
1638	Publishes his *Discorsi e dimostrazioni matematiche intorno a due nuove scienze* (*Discourses and Demonstrations Concerning Two New Sciences*), which is the beginning of modern mechanics.
1642	Galileo dies on January 8.

Introduction

Having discussed Aristotle's mechanics and the mathematics of motion, we turn to the very important contributions made in this area by the great Italian physicist Galileo Galilei (1564–1642). Galileo's main contributions to mechanics appear in his *Dialogue Concerning the Two Chief World Systems—Ptolemaic and Copernican* (1632) and in a fuller, more mathematical form in his *Discourses and Demonstrations Concerning Two New Sciences* (1638).

In studying Galileo, we shall seek answers to the following questions.
1. How do bodies move on the Earth? Galileo's primary concern was with bodies in free fall, but he also investigated pendula, balls rolling down inclined planes, and projectile motion.
2. Did Galileo have a notion of inertia and did he discover the law of inertia?
3. What was the relation between Galileo's mechanics and the Copernican system?

Allied to these scientific questions are a number of historical and/or philosophical questions.
1. What role did experiment play in Galileo's methodology?
2. To what extent was Galileo's thought indebted to the ideas of his predecessors?
3. What overall philosophical position did he espouse?
4. In general, how significant was his contribution to mechanics?

Does a Falling Body's Weight Influence Its Rate of Fall?

It might seem that the motion of bodies near the Earth's surface would be easy to analyze. Such is not the case. For example, a weight dropped from a 100 foot tower falls to the Earth in only 2.5 seconds. Moreover, friction, air resistance, etc. further impede observation. Part of Galileo's brilliance was his ability to find ways of relating observable results to this difficult problem. Early in his *Two New Sciences,* he discusses whether bodies fall at rates related to their weight, as Aristotle and many of his successors had claimed. Aristotle had also denied the possibility of a vacuum, this conclusion deriving from his claim that the fall of bodies in a medium is inversely proportional to the density of the medium.[1] Because the density of a vacuum is zero, the speed of a body moving in that vacuum becomes infinite, which Aristotle thought impossible.

---

[1] See the discussion in Chapter One, pp. 7–11.

In reading the following selection from Galileo's *Two New Sciences*, it is important to note that Galileo wrote both this work and his *Two World Systems* in a dialogue format employing three speakers: Simplicio, Salviati, and Sagredo. Simplicio (named after Simplicius, an ancient commentator on Aristotle's writings) defends Aristotelianism, whereas Salviati presents Galileo's ideas. Sagredo, who appears as a supposedly neutral intermediary, typically agrees with Salviati. Let us now turn to Galileo's *Two New Sciences*.

Galileo, *Two New Sciences*, trans. Drake, pp. 65–69[2]

Sagr.   In the denial of interpenetration I am completely on the side of the Peripatetic philosophers. With regard to the void, I should like to hear judiciously weighed the demonstration with which Aristotle refutes it, and that with which you, [Salviati], oppose him. Do me the favor, Simplicio, of providing Aristotle's proof exactly; and you, Salviati, shall reply to it.

Simp.   As I recall it, Aristotle does battle against some ancients who introduced the void as necessary for motion, saying that no motion could exist without it. Aristotle, opposing this [view], proves that on the contrary, the occurrence of motion, which we see, destroys the supposition of the void; and these are his steps. He makes two assumptions; one concerning moveables differing in heaviness but moving in the same medium, and the other concerning a given moveable moved in different mediums. As to the first, he assumes that moveables differing in heaviness are moved in the same medium with unequal speeds, which maintain to one another the same ratio as their weights [*gravità*]. Thus, for example, a moveable ten times as heavy as another, is moved ten times as fast. In the other supposition he takes it that the speeds of the same moveable through different mediums are in inverse ratio to the crassitudes or densities of the mediums. Assuming, for example, that the crassitude of water is ten times that of air, he would have it that the speed in air is ten times the speed in water.

From this second supposition he derives his proof [against the void] in this form: Since the tenuity of the void exceeds by an infinite interval the corpulence, though most rare [*sottilissima*], of any filled

---

[2]Galileo Galilei, *Two New Sciences, Including Centers of Gravity and Force of Percussion*, trans. by Stillman Drake (University of Wisconsin Press, 1974). The contents of all footnotes to the selections from Galileo in this chapter are due to the author, not Galileo or the translator.

medium [*mezzo pieno*], every moveable that is moved through some space in some time through a filled medium must be moved through the void in a single instant; but for motion to be made instantaneously is impossible; therefore, thanks to motion, the void is impossible.

Salv.   This argument is seen to be *ad hominem*; that is, it goes against those who would have the void as necessary for motion. Hence if I accept the argument as conclusive and grant that motion does not take place in the void, the supposition of the void taken absolutely, and not just in relation to motion, is not thereby destroyed.

But to say what those ancients [attacked by Aristotle] would perhaps reply, so that we may better judge the conclusiveness of Aristotle's argument, I think it possible to go against his assumptions and deny both of them. As to the first one, I seriously doubt that Aristotle ever tested [*sperimentasse*] whether it is true that two stones, one ten times as heavy as the other, both released at the same instant to fall from a height, say, of one hundred braccia, differed so much in their speeds that upon the arrival of the larger stone upon the ground, the other would be found to have descended no more than [*né anco*] ten braccia.

Simp.   But it is seen from his words that he appears to have tested this, for he says "We see the heavier...." Now this "We see" suggests that he had made the experiment [*fatta l'esperienza*].

Sagr.   But I, Simplicio, who have made the test, assure you that a cannonball that weighs one hundred pounds (or two hundred, or even more) does not anticipate by even one span the arrival on the ground of a musket ball of no more than half [an ounce], both coming from a height of two hundred braccia.

Salv.   But without other experiences, by a short and conclusive demonstration, we can prove clearly that it is not true that a heavier moveable is moved more swiftly than another, less heavy, these being of the same material, and in a word, those of which Aristotle speaks. Tell me, Simplicio, whether you assume that for every heavy falling body there is a speed determined by nature such that this cannot be increased or diminished except by using force or opposing some impediment to it.

Simp.   There can be no doubt that a given moveable in a given medium has an established speed determined by nature, which cannot be increased except by conferring on it some new impetus, nor diminished save by some impediment that retards it.

Salv.   Then if we had two moveables whose natural speeds were unequal, it is evident that were we to connect the slower to the faster, the latter would be partly retarded by the slower, and this would be partly speeded up by the faster. Do you not agree with me in this opinion?

Simp.   It seems to me that this would undoubtedly follow.

Salv.   But if this is so, and if it is also true that a large stone is moved with eight degrees of speed, for example, and a smaller one with four [degrees], then joining both together, their composite will be moved with a speed less than eight degrees. But the two stones joined together make a larger stone than that first one which was moved with eight degrees of speed; therefore this greater stone is moved less swiftly than the lesser one. But this is contrary to your assumption. So you see how, from the supposition that the heavier body is moved more swiftly than the less heavy, I conclude that the heavier moves less swiftly.

Simp.   I find myself in a tangle, because it still appears to me that the smaller stone added to the larger adds weight to it; and by adding weight, I don't see why it should not add speed to it, or at least not diminish this [speed] in it.

Salv.   Here you commit another error, Simplicio, because it is not true that the smaller stone adds [*accresca*] weight to the larger.

Simp.   Well, that indeed is beyond my comprehension.

Salv.   It will not be beyond it a bit, when I have made you see the equivocation in which you are floundering. Note that one must distinguish heavy bodies put in motion from the same bodies in a state of rest. A large stone placed in a balance acquires weight with the placement on it of another stone, and not only that, but even the addition of a coil of hemp will make it weigh more by the six or seven ounces that the hemp weighs. But if you let the stone fall freely from a height with the hemp tied to it, do you believe that in this motion the hemp would weigh on the stone, and thus necessarily speed up its motion? Or do you believe it would retard this by partly sustaining the stone?

We feel weight on our shoulders when we try to oppose the motion that the burdening weight would make; but if we descended with the same speed with which such a heavy body would naturally fall, how would you have it press and weigh on us? Do you not see that this would be like trying to lance someone who was running ahead with as much speed as that of his pursuer, or more? Infer, then, that in free and natural fall the smaller stone does not weigh upon the larger, and hence does not increase the weight as it does at rest.

Simp.   But what if the larger [stone] were placed on the smaller?

Salv.   It would increase the weight if its motion were faster. But it was already concluded that if the smaller were slower, it would partly retard the speed of the larger so that their composite, though larger than before, would be moved less swiftly, which is against your assumption. From this we conclude that both great and small bodies, of the same specific gravity, are moved with like speeds.

Simp.   Truly, your reasoning goes along very smoothly; yet I find it hard to believe that a birdshot must move as swiftly as a cannonball.

Salv.   You should say "a grain of sand as [fast as] a millstone." But I don't want you, Simplicio, to do what many others do, and divert the argument from its principal purpose, attacking something I said that departs by a hair from the truth, and then trying to hide under this hair another's fault that is a big as a ship's hawser. Aristotle says, "A hundred-pound iron ball falling from the height of a hundred braccia hits the ground before one of just one pound has descended a single braccio." I say that they arrive at the same time. You find, on making the experiment, that the larger anticipates the smaller by two inches; that is, when the larger one strikes the ground, the other is two inches behind it. And now you want to hide, behind those two inches, the ninety-nine braccia of Aristotle, and speaking only of my tiny error, remain silent about his enormous one.

Aristotle declares that moveables of different weight are moved (to the extent this depends on heaviness) through the same medium with speeds proportional to their weights. He gives as an example moveables in which the pure and absolute effect of weight can be discerned, leaving aside those other considerations of shapes and of certain very tiny forces [*momenta*], which introduce great changes [*alterazione*] from the medium, and which alter the simple effect of heaviness alone. Thus one sees gold, which is most heavy, more so than any other material, reduced to a very thin leaf that goes floating through the air, as do rocks crushed into fine dust. If you wish to maintain your general proposition, you must show that the ratio of speeds is observed in all heavy bodies, and that a rock of twenty pounds is moved ten times as fast as a two-pound rock. I say this is false, and that in falling from a height of fifty or a hundred braccia, they will strike the ground at the same moment.

Simp.   Perhaps from very great heights, of thousands of braccia, that would follow which is not seen at these lesser heights.

Salv.   If that is what Aristotle meant, you saddle him with a further error that would be a lie. For no such vertical heights are found on earth, so it is clear that Aristotle could not have made that trial; yet you want to persuade us that he did so because he says that the effect "is seen."

Commentary on Galileo's *Two New Sciences*, pp. 65–69

This section of Galileo's book is rather remarkable. We find him challenging Aristotle's doctrine on the basis of an experiment, but not at all of the type that one might expect. The experiment has the form of a "thought experiment." In particular, Galileo in effect says: let us assume that Aristotle's claim is correct that a body's rate of fall depends on its weight and let us apply this claim to a case involving two weights, one ten times heavier than the other. On Aristotle's analysis, the heavier weight ought to fall ten times faster than the light one. But, says Galileo, consider what this analysis would entail for the case of two weights tied together. On the one hand, the lighter weight should impede the heavier, making the combined weights fall more slowly than the heavier weight alone. On the other hand, we have created a weight heavier than the original weights and hence it should fall faster than either of them. Galileo's point is that a claim that entails mutually contradictory conclusions must itself be erroneous. Thus we see that physical experiment played no role in this part of Galileo's argument.

In the final paragraph of the passage from Galileo, he faults Aristotle for claiming to have witnessed an experiment he could not have made. Leaving aside the question of the legitimacy of this complaint, we can ask about the experiments that Sagredo claims that Galileo performed. These were such that in a fall of about 100 braccia (about 200 feet), two weights of different heaviness were found to reach the ground differing in position by only "two inches." This claim is surely excessive. We can calculate (using modern methods) that each body, if falling from this distance, would be moving at the speed of about 110 feet per second at the time it strikes the ground. Given such rapid motion and the difficulty of telling whether the two bodies were released simultaneously, we can suspect that Galileo's claim goes far beyond what it would have been possible for him to observe. It is interesting in this context to note that no direct documentation from the time of Galileo's life supports the widespread belief that he dropped weights from the leaning tower of Pisa.

We can approach the same point—Galileo's relation to experimentation—from another direction. Some Galileo manuscripts from early in his career recount experiments in which he compared what happened when he dropped a wooden and iron ball from the same substantial elevation. What he found was that after he had released the two balls simultaneously from his hands, the wooden ball in the early part of its descent led the way, but that the iron ball then overtook it. Modern scholars, knowing that theory dictated that in a vacuum, the balls should fall identically but also that for fall in a medium such as the air, the iron ball should more successfully overcome friction and reach the Earth slightly ahead of the wooden ball, were puzzled by Galileo's

reported experimental results. Nonetheless, when experiments were performed, they showed that Galileo had correctly reported what happens—the
wooden ball at first leads, but then trails! Why this should be was finally explained partly with the aid of photographs. It was learned that when we hold
and then simultaneously (we hope) release a heavy and light ball, we in fact
almost invariably release the lighter ball sooner, giving it a small head start.
This is due to differential muscle fatigue in our hands. What one sees in this
is that it is very easy to exaggerate the role of observation and experiment
in the early history of mechanics.[3]

Galileo on the Motion of Pendula

In the next section of Galileo's *Two New Sciences,* he discusses how media
resist falling bodies. In this section, he cites various arguments to show
that the effect of the medium does not manifest itself as a denominator as
Aristotle in effect assumed, but rather is subtractive. Thus the medium exerts
a force upward, which should be subtracted from the downward force of
the body. This frees Galileo from a number of problems in the Aristotelian
system, not least the problem present in that system of why some bodies
(e.g., wood balls) rise rather than sink in water.

Galileo had long been interested in the motion of pendula, i.e., weights
suspended on strings. In fact, one of his earliest discoveries was of the
isochronism of the pendulum. That the motion of pendula is not unrelated
to the problem of falling bodies becomes clear from the following section of
Galileo's *Two New Sciences.*

Galileo, *Two New Sciences*, trans. Drake, pp. 86–91

Salv. What I have set forth thus far is new; especially that no difference of weight, however great, plays any part at all in diversifying the
speeds of moveables, so that as far as speed depends on weight, all
moveables are moved with equal celerity. At first glance, this seems
so remote from probability that, if I did not have some way of eluci-
dating it and making it clear as daylight, it would have been better to
remain silent than to assert it. So now that it has escaped my lips, I
must not neglect any experiment or reason that can corroborate it.

---

[3]On these experiments, see I. B. Cohen, *The Birth of the New Physics,* rev. ed. (W. W. Norton,
1985), pp. 194–195 and Michael Sharratt, *Galileo: Decisive Innovator* (Cambridge University Press, 1994), pp. 49–51.

Sagr. Not only this proposition, but many others of yours are so far from the opinions and teachings commonly accepted, that to broadcast them publicly will excite against them a great number of contradictors; for the innate condition of men is to look askance on others working in their field whose studies reveal truth or falsity which they themselves fail to perceive. By calling such men [as you] "innovators of doctrines," a title most unpleasant to the ears of the multitude, they strain to cut those knots they cannot untie, and to demolish with underground mines those edifices which have been built by patient artificers, working with ordinary instruments. But to us, who are far from any such motives, the experiments and reasons adduced up to this point are quite satisfactory.

Salv. The experiment made with two moveables, as different as possible in weight, made to fall from a height in order to observe whether they are of equal speed, labors under certain difficulties. If the height is very great, the medium that must be opened and driven aside by the impetus of the falling body will be of greater prejudice to the small momentum of a very light moveable than to the force of a very heavy one, and over a long distance the light one will remain behind. But in a small height it may be doubtful whether there is really no difference [in speeds], or whether there is a difference but it is unobservable. So I fell to thinking how one might many times repeat descents from small heights, and accumulate many of those minimal differences of time that might intervene between the arrival of the heavy body at the terminus and that of the light one, so that added together in this way they would make up a time not only observable, but easily observable.

In order to make use of motions as slow as possible, in which resistance by the medium does less to alter the effect dependent upon simple heaviness, I also thought of making the moveables descend along an inclined plane not much raised above the horizontal. On this, no less than along the vertical, one may observe what is done by heavy bodies differing in weight. Going further, I wanted to be free of any hindrance that might arise from contact of these moveables with the said tilted plane. Ultimately, I took two balls, one of lead and one of cork, the former being at least a hundred times as heavy as the latter, and I attached them to equal thin strings four or five braccia long, tied high above. Removed from the vertical, these were set going at the same moment, and falling along the circumferences of the circles described by the equal strings that were the radii, they passed the vertical and returned by the same path. Repeating their goings and comings a good hundred times by themselves, they sensibly showed that the heavy one kept time with the light one so well that not in a hundred oscillations, nor in a thousand, does it get ahead in time even by a moment, but the two travel with equal pace. The operation of the medium is also perceived; offering some impediment to the motion, it diminishes the oscillations of the cork much more than those of the lead. But it does not make them more frequent, or less so; indeed, when the arcs passed by the cork were not more than five or six degrees, and those of the lead were fifty or sixty, they were passed over in the same times.

Simp. If that is so, why then will the speed of the lead not be [called] greater than that of the cork, seeing that it travels sixty degrees in the time that the cork hardly passes six?

Salv. And what would you say, Simplicio, if both took the same time in their travels when the cork, removed thirty degrees from the vertical, had to pass an arc of sixty, and the lead, drawn but two degrees from the same point, ran through an arc of four? Would not the cork then be as much the faster? Yet experience shows this to happen. But note that if the lead pendulum is drawn, say, fifty degrees from the vertical and released, it passes beyond the vertical and runs almost another fifty, describing an arc of nearly one hundred degrees. Returning of itself, it describes another slightly smaller arc; and continuing its oscillations, after a great number of these it is finally reduced to rest. Each of those vibrations is made in equal times, as well that of ninety degrees as that of fifty, or twenty, or ten, or of four. Consequently the speed of the moveable is always languishing, since in equal times it passes successively arcs ever smaller and smaller. A similar effect, indeed the same, is produced by the cork that hangs from another thread of equal length, except that this comes to rest in a smaller number of oscillations, as less suited by reason of its lightness to overcome the impediment of the air. Nevertheless, all its vibrations, large and small, are made in times equal among themselves, and also equal to the times of the vibrations of the lead. Whence it is true that if, while the lead passes over an arc of fifty degrees, the cork passes over only ten, then the cork is slower than the lead; but it also happens in reverse that the cork passes along the arc of fifty while the lead passes that of ten, or six; and thus, at different times, the lead will now be faster, and again the cork. But if the same moveables also pass equal arcs in the same equal times, surely one may say that their speeds are equal.

Simp. This reasoning seems to me conclusive, and also it seems it isn't; my mind feels a kind of confusion that rises from the moving of both moveables now quickly, now slowly, and again extremely slowly, so that I can't get it straight in my head whether it is true that their speeds are always equal.

Sagr. I'd like to say a word, Salviati. Tell me, Simplicio, whether you grant that it may be said with absolute truth that the speed of the cork and that of the lead are equal every time that they both start at the same moment from rest and, moving along the same slopes, always pass equal spaces in equal times.

Simp.   In this there is no room for doubt; it cannot be contradicted.

Sagr. Now it happens with either pendulum that it passes now sixty degrees, now thirty, now ten, now eight, now four, now two, and so on. And when both pass the arc of sixty degrees, they pass this in the same time; in the arc of fifty, both bodies spend the same time; so in the arc of thirty, of ten, and the rest. Thus it is concluded that the speed of the lead in the arc of sixty degrees is equal to the speed of the cork in the same arc of sixty; and that the speeds in the arc of fifty are still equal to each other, and so on in the rest. But nobody says that the speed employed in the arc of sixty [degrees] is equal to that consumed in the arc of fifty, nor this speed to that in the arc of thirty, and so on. The speeds are always less in the smaller arcs, which we deduce by seeing with our own eyes that the same body spends as much time in passing the large arc of sixty degrees as in passing the smaller of fifty or the very small arc of ten; and in sum, that all arcs are passed in equal times. It is therefore true that the lead and the cork do go retarding their motion according to the diminution of the arcs, but their agreement in maintaining equality of speed in every arc that is passed by both of them remains unaltered.

I wanted to say this to learn whether I have correctly understood Salviati's idea, rather than because I believe that Simplicio deserved a clearer explanation than that of Salviati, who here, as in all things, is most lucid. Usually he unravels questions that seem not only obscure, but repugnant to nature and the truth, [and does this] by reasons, observations, or experiences that are well known and familiar to everyone. I have heard from various people that this has given occasion to a certain highly esteemed professor to deprecate his discoveries [*novità*], holding them to be base, as depending on foundations too low and common—as if it were not the most admirable and estimable condition of the demonstrative sciences that they arise and flow from well-known principles, understood and conceded by all.

But let us go on feasting on these light foods, assuming that Simplicio is now willing to assume and grant that the internal heaviness of different moveables has no part at all in diversifying their speeds, so that all, so far as weight is concerned, move with the same speeds.

Salviati, tell us how you explain the sensible and obvious inequalities of motion, and reply to Simplicio's objection, which I also confirm, that we see a cannonball fall more swiftly than a lead shot, whereas according to you, the difference of speed will be small. I counter this with some moveables of the same material, of which the larger will fall in less than a pulsebeat, in one medium, through a space that others smaller will not pass in an hour, or four, or twenty. These are stones, and fine sand, to say nothing of that dust that muddies water, a medium in which this does not fall through two braccia in many hours, through which [distance] rocks, and not very big ones at that, fall in a pulsebeat.

Salv.   The part played by the medium in more greatly retarding moveables according as they are less in specific gravity has already been explained; this occurs by the subtraction of weight. How a given medium can reduce speed very differently in bodies that differ only in size, and are of the same material and shape, requires for its explanation subtler reasoning than that which suffices to understand how a flat shape in a moveable, or motion of the medium against one, retards its speed.

For the present problem, I reduce the reason to the roughness and porosity found commonly, and for the most part necessarily, at the surface of solid bodies. In motion, those irregularities strike the air or other surrounding medium, an evident sign of which is that we hear bodies hum when they fly rapidly through the air, even when rounded as thoroughly as possible; and they not only hum, but they are heard to whistle and hiss if some notable cavity or protuberance exists in them. It is also seen that every round solid turned on a lathe makes a little breeze. Again, we hear a humming, very high in pitch, made by a top when it spins rapidly on the ground. The pitch of this tone deepens as the spinning languishes bit by bit; this also necessarily argues hindrances by the air of the surface roughnesses, however tiny. It cannot be doubted that in the descent of moveables these [irregularities] rub against the surrounding fluid and bring about retardation of speed, greater as the surface is larger, as is the case with smaller solids in comparison with larger ones.

Commentary on Galileo's *Two New Sciences*, pp. 86–91

These pages recount Galileo's discovery of the **isochronism of the pendulum**. What he reports is that the time of swing of any pendulum is the same, no matter what the weight or material of the pendulum bob may be and no matter what the size of the arc through which it swings is. The length of the cord does, however, make a difference, as Galileo soon specifies. Moreover, he uses this evidence to support his theory that bodies fall at the same rate independently of their weight. This discovery was important in yet another way: it suggested the possibility of the pendulum clock. Because of the isochronism of the pendulum, it became possible to make pendulum driven clocks that kept time with greater accuracy than that attainable before this time.

Galileo's final argument in this section is very interesting. He explains why pendulum bobs of identical materials but different sizes gradually move through different arcs, although retaining the same period. In particular, if a large and a small sphere of iron are set swinging through identical arcs, the larger sphere will retain its motion the better, i.e., it will swing through a larger arc than the smaller one. This seems to contradict his discovery of the isochronism of the pendulum. His response is quite ingenious and depends on mathematical considerations. He assumes that the two spheres are identical in shape and that each has a certain number of "irregularities" or roughnesses per unit of surface area. Then he turns to mathematics. Putting his argument into a slightly more modern dress, we know that the surface area of a sphere of radius R is $4\pi R^2$, whereas its volume is $\frac{4}{3}\pi R^3$. Let us use these formulas to compare two spheres, the first of radius 1, the second of radius 2. We see that the surface area of a sphere of radius 1 equals 4π, whereas its volume is $\frac{4}{3}\pi$, making the ratio of surface area to volume equal 3. For a sphere of radius 2, its area is 16π and its volume is $\frac{4}{3}\pi \cdot 8 = \frac{32}{3}\pi$. The ratio of its surface to its volume is $\frac{3}{2}$. Thus we see that as spheres grow larger, the ratios of their surface areas to volume constantly decreases. In other words, larger spheres show fewer "irregularities" per unit of weight than do smaller ones. Because, according to Galileo, it is the "irregularities" that cause the object to swing through shorter arcs, the smaller body suffers more retardation. This argument can also be used to explain why in free fall, larger bodies strike the ground slightly sooner than smaller bodies of the same specific gravity and shape.

A few pages after this, Galileo goes on to quantify, at least to some extent, the relationship between the length of a pendulum string and its period of swing. He writes:

> As to the ratio of times of oscillation of bodies hanging from strings of different lengths, those times are as the square roots of the string lengths; or we should say that the lengths are as the doubled ratios, or squares, of the times. Thus if, for example, you want the time of oscillation of one pendulum to be double the time of another, the length of its string must be four times that of the other; or if in the time of one vibration of the first, another is to make three, then the string of the first will be nine times as long as that of the other. It follows from this

that the lengths of the strings have to one another the [inverse] ratio of the squares of the number of vibrations made in a given time.[4]

We can write Galileo's result as an equation in which T = period of swing and L = length of the cord:

$$\frac{T_1}{T_2} = \frac{\sqrt{L_1}}{\sqrt{L_2}}.$$

Galileo's analysis makes it possible to solve the following problems.

Problems

1. Consider two pendula P_1 and P_2, P_1 consisting of 6 oz. of brass and P_2 of 3 oz. of lead. If P_1 when suspended on a cord of length L_1 swings through and returns along an arc of $40°$ in 2 seconds, how long will P_2 take to swing through $60°$ (and return) if its cord is of length $4L_1$?

2. Suppose you see a pendulum (P_1) of length L_1 swinging through a small arc in 1 second. Another longer pendulum (P_2) is positioned so that you cannot see its lower end, but you observe that its period is 3 seconds. Calculate its length in terms of L_1.

3. Salviati delights Sagredo by the following description of three pendula in motion.

> *Salv.* Seeing that you like these novelties so well, I must show you how the eye, too, and not just the hearing, can be amused by seeing the same play that the ear hears.
>
> Hang lead balls, or similar heavy bodies, from three threads of different lengths, so that in the time that the longest makes two oscillations, the shortest makes four and the other makes three. This will happen when the longest contains sixteen spans, or other units, of which the middle [length] contains nine, and the smallest, four. Removing all these from the vertical and releasing them, an interesting interlacing of the threads will be seen, with varied meetings such that at every fourth oscillation of the longest, all three arrive unitedly at the same terminus; and from this they depart, repeating again the same period. The mixture of oscillations is such that when made by [tuned] strings, it renders to the hearing an octave with the intermediate fifth. And if with the same arrangements we modify the lengths of other strings so

---

[4]Galileo Galilei, *Two New Sciences, Including Centers of Gravity and Force of Percussion,* trans. Stillman Drake (University of Wisconsin Press, 1974), pp. 97–98.

that their vibrations answer to those of other musical intervals which are consonances, other interlacings will be seen in which, at the determinate times and after definite numbers of vibrations, all the strings (let them be three or four) agree in coming at the same moment to the terminus of their oscillations, and begin from there another like period.[5]

Use the previously given formula to check Galileo's result.

Galileo on Accelerated Motion and Free Fall

Galileo's *Two New Sciences* presents his thoughts on two sciences: that of the strength of materials and that dealing with the motion of bodies. Whereas the first two of the four days of the dialogue focus on the strength of materials, the final two days contain his detailed analysis of terrestrial motion. At the beginning of the third day, he reveals what new results concerning motion he has attained.

Galileo, *Two New Sciences*, trans. Drake, pp. 147–148

THIRD DAY

[*Salviati* (reading from Galileo's Latin treatise):]

On Local Motion

We bring forward [promovemus] *a brand new science concerning a very old subject.*

There is perhaps nothing in nature older than MOTION, *about which volumes neither few nor small have been written by philosophers; yet I find many essentials* [symptomata] *of it that are worth knowing which have not even been remarked, let alone demonstrated. Certain commonplaces have been noted, as for example that in natural motion, heavy falling things continually accelerate; but the proportion according to which this acceleration takes place has not yet been set forth. Indeed no one, so far as I know, has demonstrated that the spaces run through in equal times by a moveable descending from rest maintain among themselves the same rule* [rationem] *as do the odd numbers following upon unity. It has been observed that missiles or projectiles trace out a line somehow curved, but no one has brought out that this is a parabola. That it is, and other things neither few nor less*

---

[5]Galileo, *Two New Sciences*, p. 107.

worthy [than this] of being known, will be demonstrated by me, and (what is in my opinion more worthwhile) there will be opened a gateway and a road to a large and excellent science of which these labors of ours shall be the elements, [a science] into which minds more piercing than mine shall penetrate to recesses still deeper.

We shall divide this treatise into three parts. In the first part we consider that which relates to equable or uniform motion; in the second, we write of motion naturally accelerated; and in the third, of violent motion, or of projectiles.

Commentary on Galileo's *Two New Sciences*, pp. 147–148

Following that introduction are six pages in which Galileo sets out his definition of "**uniform motion**" and develops six elementary theorems concerning it. Galileo's definition of uniform motion is: "*Equal or uniform motion I understand to be that of which the parts run through by the moveable in any equal times whatever are equal to one another.*"[6] This is what we would call motion with constant speed. The first of his theorems is: "*If a moveable equably carried [latum] with the same speed passes through two spaces, the times of motion will be to one another as the spaces passed through.*"[7] Galileo's presentation of this is almost entirely within a geometrical context, as will be evident in the selections that follow. We can, however, translate his first theorem into an algebraic equation; for a body moving at constant speed, then for s symbolizing distance and t representing time,

$$\frac{s_1}{s_2} = \frac{t_1}{t_2}.$$

Galileo's second theorem is: "*If a moveable passes through two spaces in equal times, these spaces will be to one another as the speeds. And if the spaces are as the speeds, the times will be equal.*"[8] Translated into symbolic form with v representing speed, this becomes:

$$\text{If } t_1 = t_2, \text{ then } \frac{s_1}{s_2} = \frac{v_1}{v_2}, \text{ and if } \frac{s_1}{s_2} = \frac{v_1}{v_2}, \text{ then } t_1 = t_2.$$

The pattern apparent in the first two theorems is evident in the four remaining theorems: each can be derived as a special case from the basic formula that $v = \frac{s}{t}$. Consequently, we need not follow his discussion in detail, but can instead proceed to more complex matters.

---

[6]Galileo, *Two New Sciences*, p. 148.
[7]Galileo, *Two New Sciences*, p. 149.
[8]Galileo, *Two New Sciences*, p. 150.

These appear in the next section, which is headed "On Naturally Accelerated Motion." The selection that follows is from the first part of that section. In it, Galileo treats "**uniform acceleration**," a property he attributes to freely falling bodies. The problem he confronts is how bodies move when falling from a height. He is convinced that they accelerate; what he attempts to determine is **how** they accelerate. His initial assumption is that they accelerate in some mathematically simple way. Two possibilities are discussed: (1) that their speed increases in direct proportion to distance, and (2) that speed is proportional to time of descent. To put this question in the form of an equation: letting v_f represent the final speed of a body moving in free fall, is $v_f \propto s$ or is $v_f \propto t$? (Note: $\propto$ is a symbol meaning "proportional to.") Of course, neither of these may be correct, but the mathematics of the problem will be simplest if one or the other turns out to be the case. Galileo had for some time believed the first proportion to be the true one. He presented this claim in a manuscript, *De Motu,* which Galileo composed around 1590. By 1604, however, he had attained the correct law, although his derivation remained deficient for some time.

Galileo had to face the further problem of proving that the $v \propto t$ assumption actually applies to the real world. Near the end of the selection, he indicates his first method of doing this. In reading these pages, in which Salviati continues to read from Galileo's treatise, it is interesting to observe the philosophical approach taken by the famous scientist. Finally, it is best to avoid getting bogged down in detail; a good technique will be to read the materials quickly, looking for the general direction of the argument. Then, proceed more carefully through them, trying to decide what is crucial to the argument and what is peripheral. The commentary that follows them should help with this.

Two New Sciences, trans. Drake, pp. 153–170

On Naturally Accelerated Motion

Those things that happen which relate to equable motion have been considered in the preceding book; next, accelerated motion is to be treated of.

And first, it is appropriate to seek out and clarify the definition that best agrees with that [accelerated motion] which nature employs. Not that there is anything wrong with inventing at pleasure some kind of motion and theorizing about its consequent properties, in the way that some men have derived spiral and conchoidal lines from certain motions, though nature makes no use of these [paths]; and by pretending these, men have laudably demonstrated their essentials from assumptions [ex suppositione].

But since nature does employ a certain kind of acceleration for descending heavy things, we decided to look into their properties so that we might be sure that the definition of accelerated motion which we are about to adduce agrees with the essence of naturally accelerated motion. And at length, after continual agitation of mind, we are confident that this has been found, chiefly for the very powerful reason that the essentials successively demonstrated by us correspond to, and are seen to be in agreement with, that which physical experiments [naturalia experimenta] show forth to the senses. Further, it is as though we have been led by the hand to the investigation of naturally accelerated motion by consideration of the custom and procedure of nature herself in all her other works, in the performance of which she habitually employs the first, simplest, and easiest means. And indeed, no one of judgment believes that swimming or flying can be accomplished in a simpler or easier way than that which fish and birds employ by natural instinct.

Thus when I consider that a stone, falling from rest at some height, successively acquires new increments of speed, why should I not believe that those additions are made by the simplest and most evident rule? For if we look into this attentively, we can discover no simpler addition and increase than that which is added on always in the same way. We easily understand that the closest affinity holds between time and motion, and thus equable and uniform motion is defined through uniformities of times and spaces; and indeed, we call movement equable when in equal times equal spaces are traversed. And by this same equality of parts of time, we can perceive the increase of swiftness to be made simply, conceiving mentally that this motion is uniformly and continually accelerated in the same way whenever, in any equal times, equal additions of swiftness are added on.

Thus, taking any equal particles of time whatever, from the first instant in which the moveable departs from rest and descent is begun, the degree of swiftness acquired in the first and second little parts of time [together] is double the degree that the moveable acquired in the first little part [of time]; and the degree that it gets in three little parts of time is triple; and in four, quadruple that same degree [acquired] in the first particle of time. So, for clearer understanding, if the moveable were to continue its motion at the degree of momentum of speed acquired in the first little part of time, and were to extend its motion successively and equably with that degree, this movement would be twice as slow as [that] at the degree of speed obtained in two little parts of time. And thus it is seen that we shall not depart far from the correct rule if we assume that intensification of speed is made according to the extension of time; from which the definition of the motion of which we are going to treat may be put thus:

[DEFINITION]

I say that that motion is equably or uniformly accelerated which, abandoning rest, adds on to itself equal momenta of swiftness in equal times.

Sagr. Just as it would be unreasonable for me to oppose this, or any other definition whatever assigned by any author, all [definitions] being arbitrary, so I may, without offence, doubt whether this definition, conceived and assumed in the abstract, is adapted to, suitable for, and verified in the kind of accelerated motion that heavy bodies in fact employ in falling naturally. And since it seems that the Author promises us that what he has defined is the natural motion of heavy bodies, I should like to hear you remove certain doubts that disturb my mind, so that I can then apply myself with better attention to the propositions that are expected, and their demonstrations.

Salv. It will be good for you and Simplicio to propound the difficulties, which I imagine will be the same ones that occurred to me when I first saw this treatise, and that our Author himself put to rest for me in our discussions, or that I removed for myself by thinking them out.

Sagr. I picture to myself a heavy body falling. It leaves from rest; that is, from the deprivation of any speed whatever, and enters into motion in which it goes accelerating according to the ratio of increase of time from its first instant of motion. It will have obtained, for example, eight degrees of speed in eight pulse-beats, of which at the fourth beat it will have gained four; at the second [beat], two; and at the first, one. Now, time being infinitely divisible, what follows from this? The speed being always diminished in this ratio, there will be no degree of speed, however small (or we might say, "no degree of slowness, however great"), such that the moveable will not be found to have this [at some time] after its departure from infinite slowness, that is, from rest. Thus if the degree of speed that it had at four beats of time were such that, maintaining this uniformly, it would run two miles in one hour, while with the degree of speed that it had at the second beat it would have made one mile an hour, it must be said that in instants of time closer and closer to the first [instant] of its moving from rest, it would be found to be so slow that, continuing to move with this slowness, it would not pass a mile in an hour, nor in a day, nor in a year, nor in a thousand [years], and it would not pass even one span in some still longer time. Such events I find very hard to accommodate in my imagination, when our senses show us that a heavy body in falling arrives immediately at a very great speed.

Salv. This is one of the difficulties that gave me pause at the outset; but not long afterward I removed it, and its removal was effected by the same experience that presently sustains it for you.

You say that it appears to you that experience shows the heavy body, having hardly left from rest, entering into a very considerable speed; and I say that this same experience makes it clear to us that the first impetuses of the falling body, however heavy it may be, are very slow indeed. Place a heavy body on some yielding material, and leave it until it has pressed as much as it can with its mere weight. It is obvious that if you now raise it one or two braccia, and then let it fall on the same material, it will make a new pressure on impact, greater than it made by its weight alone. This effect will be caused by the falling moveable in conjunction with the speed gained in fall, and will be greater and greater according as the height is greater from which the impact is made; that is, according as the speed of the striking body is greater. The amount of speed of a falling body, then, we can estimate without error from the quality and quantity of its impact.

But tell me, gentlemen: if you let a sledge fall on a pole from a height of four braccia, and it drives this, say, four inches into the ground, and will drive it much less from a height of two braccia, and still less from a height of one, and less yet from a span only; if finally it is raised but a single inch, how much more will it accomplish than if it were placed on top [of the pole] without striking it at all? Certainly very little. And its effect would be quite imperceptible if it were lifted only the thickness of a leaf. Now, since the effect of impact is governed by the speed of a given percussent, who can doubt that its motion is very slow and minimal when its action is imperceptible? You now see how great is the force of truth, when the same experience that seemed to prove one thing at first glance assures us of the contrary when it is better considered.

But without restricting ourselves to this experience, though no doubt it is quite conclusive, it seems to me not difficult to penetrate this truth by simple reasoning. We have a heavy stone, held in the air at rest. It is freed from support and set at liberty; being heavier than air, it goes falling downward, not with uniform motion, but slowly at first and continually accelerated thereafter. Now, since speed may be increased or diminished *in infinitum,* what argument can persuade me that this moveable, departing from infinite slowness (which is rest), enters immediately into a speed of ten degrees rather than into one of four, or into the latter before a speed of two, or one, or one-half, or one one-hundredth? Or, in short, into all the lesser [degrees] *in infinitum?*

Please hear me out. I believe you would not hesitate to grant me that the acquisition of degrees of speed by the stone falling from the state of rest may occur in the same order as the diminution and loss of those same degrees when, driven by impelling force, the stone is hurled upward to the same height. But if that is so, I do not see how it can be supposed that in the diminution of speed in the ascending stone, consuming the whole speed, the stone can arrive at rest before passing through every degree of slowness.

Simp. But if the degrees of greater and greater tardity are infinite, it will never consume them all, and this rising heavy body will never come to rest, but will move forever while always slowing down—something that is not seen to happen.

Salv. This would be so, Simplicio, if the moveable were to hold itself for any time in each degree; but it merely passes there, without remaining beyond an instant. And since in any finite time [*tempo quanto*], however small, there are infinitely many instants, there are enough to correspond to the infinitely many degrees of diminished speed.[9] It is obvious that this rising heavy body does not persist for any finite time in any one degree of speed, for if any finite time is assigned, and if the moveable had the same degree of speed at the first instant of that time and also at the last, then it could likewise be driven upward with this latter degree [of speed] through as much space [again], just as it was carried from the first [instant] to the second; and at the same rate it would pass from the second to a third, and finally, it would continue its uniform motion *in infinitum.*

Sagr. From this reasoning, it seems to me that a very appropriate answer can be deduced for the question agitated among philosophers as to the possible cause of acceleration of the natural motion of heavy bodies. For let us consider that in the heavy body hurled upwards, the force [*virtù*] impressed upon it by the thrower is continually diminishing, and that this is the force that drives it upward as long as this remains greater than the contrary force of its heaviness; then when these two [forces] reach equilibrium, the moveable stops rising

---

[9]In this section Galileo is discussing some of the paradoxes of motion presented around 460 BCE by Zeno of Elea and widely discussed by later authors. The particular Zenonian paradox dealt with here is known as "The Race Course." The idea is that one can never complete a race course, no matter how short, because one would have first to traverse half the distance, then half of the remainder, then half of that, and so on indefinitely, so that it seems that one would have to traverse an infinite number of distances and thus also an infinite distance, which is impossible. Galileo's proposed resolution of this paradox in the paragraph noted is recognized as a major step in the efforts to overcome these difficulties.

and passes through a state of rest. Here the impressed impetus[10] is [still] not annihilated, but merely that excess has been consumed that it previously had over the heaviness of the moveable, by which [excess] it prevailed over this [heaviness] and drove [the body] upward. The diminutions of this alien impetus then continuing, and in consequence the advantage passing over to the side of the heaviness, descent commences, though slowly because of the opposition of the impressed force, a good part of which still remains in the moveable. And since this continues to diminish, and comes to be overpowered in ever-greater ratio by the heaviness, the continual acceleration of the motion arises therefrom.

Simp. The idea is clever, but more subtle than sound; for if it were valid, it would explain only those natural motions which had been preceded by violent motion, in which some part of the external impetus still remained alive. But where there is no such residue, and the moveable leaves from longstanding rest, the whole argument loses its force.

Sagr. I believe you are mistaken, and that the distinction of cases made by you is superfluous, or rather, is idle. For tell me: can the thrower impress on the projectile sometimes much force, and sometimes little, so that it may be driven upward a hundred braccia, or twenty, or four, or only one?

Simp. No doubt he can.

Sagr. No less will the force impressed be able to overcome the resistance of heaviness by so little that it would not raise [the body] more than an inch. And finally, the force of projection may be so small as just to equal the resistance of the heaviness, so that the moveable is not thrown upward, but merely sustained. Thus, when you support a rock in your hand, what else are you doing but impressing on it just as much of that upward impelling force as equals the power of its heaviness to draw it downward? And do you not continue this force of yours, keeping it impressed through the whole time that you support [the rock] in your hand? Does the force perhaps diminish during the length of time that you support the rock? Now, as to this sustaining that prevents the fall of the rock, what difference does it make whether it comes from your hand, or a table, or a rope tied to it? None whatever. You must conclude, then, Simplicio, that it makes no difference at all whether the fall of the rock is preceded by a long rest, or a short

---

[10]It is important to ask various questions about what Galileo call impetus. What precisely does Galileo means by this term? Is impetus material or immaterial? How would one quantify it? Where did Galileo get the idea? And to what, if any, later idea does it lead?

one, or one only momentary, and that the rock always starts with just as much of the force contrary to its heaviness as was needed to hold it at rest.

Salv. The present does not seem to me to be an opportune time to enter into the investigation of the cause of the acceleration of natural motion, concerning which various philosophers have produced various opinions, some of them reducing this to approach to the center; others to the presence of successively less parts of the medium [remaining] to be divided; and others to a certain extrusion by the surrounding medium which, in rejoining itself behind the moveable, goes pressing and continually pushing it out. Such fantasies, and others like them, would have to be examined and resolved, with little gain. For the present, it suffices our Author that we understand him to want us to investigate and demonstrate some attributes [*passiones*] of a motion so accelerated (whatever be the cause of its acceleration) that the momenta of its speed go increasing, after its departure from rest, in that simple ratio with which the continuation of time increases, which is the same as to say that in equal times, equal additions of speed are made. And if it shall be found that the events that then shall have been demonstrated are verified in the motion of naturally falling and accelerated heavy bodies, we may deem that the definition assumed includes that motion of heavy things, and that it is true that their acceleration goes increasing as the time and the duration of motion increases.[11]

Sagr. By what I now picture to myself in my mind, it appears to me that this could perhaps be defined with greater clarity, without varying the concept, [as follows]: Uniformly accelerated motion is that in which the speed goes increasing according to the increase of space traversed. Thus for example, the degree of speed acquired by the moveable in the descent of four braccia would be double that which it had after falling through the space of two, and this would be the double of that resulting in the space of the first braccio. For there seems to me to be no doubt that the heavy body coming from a height of six braccia has, and strikes with, double the impetus that it would have from falling three braccia, and triple that which it would have from two, and six times that had in the space of one.

Salv. It is very comforting to have had such a companion in error, and I can tell you that your reasoning has in it so much of the plausible and probable, that our Author himself did not deny to me, when I

---

[11] This paragraph is very important in what it reveals about the methodological approach adopted by Galileo.

proposed it to him, that he had labored for some time under the same fallacy. But what made me marvel then was to see revealed, in a few simple words, to be not only false but impossible, two propositions which are so plausible that I have propounded them to many people, and have not found one who did not freely concede them to me.

Simp. Truly, I should be one of those who concede them. That the falling heavy body *vires acquirit eundo*[12] [acquires force in going], the speed increasing in the ratio of the space, while the momentum of the same percussent is double when it comes from double height, appear to me as propositions to be granted without repugnance of or controversy.

Salv. And yet they are as false and impossible as [it is] that motion should be made instantaneously, and here is a very clear proof of it. When speeds have the same ratio as the spaces passed or to be passed, those spaces come to be passed in equal times; if therefore the speeds with which the falling body passed the space of four braccia were the doubles of the speeds with which it passed the first two braccia, as one space is double the other space, then the time of those passages are equal; but for the same moveable to pass the four braccia and the two in the same time cannot take place except in instantaneous motion. But we see that the falling heavy body makes its motion in time, and passes the two braccia in less [time] than the four; therefore it is false that its speed increases as the space.

The other proposition is shown to be false with the same clarity. For that which strikes being the same body, the difference and momenta of the impacts must be determined only by the difference of the speeds; if therefore the percussent coming from a double height delivers a blow of double momentum, it must strike with double speed; but double speed passes the double space in the same time, and we see the time descent to be longer from the greater height.

Sagr. Too evident and too easy is this [reasoning] with which you make hidden conclusions manifest. This great facility renders the conclusions less prized than when they were under seeming contradiction. I think that people generally will little esteem ideas gained with so little trouble, in comparison with those over which long and unresolvable altercations are waged.

Salv. Things would not be so bad if men who show with great brevity and clarity the fallacies of propositions that have commonly been held to be true by people in general received only such bearable injury as scorn in place of thanks. What is truly unpleasant and

---

[12] The Drake translation erroneously reads "acquirat" here. [editor's note]

annoying is a certain other attitude that some people habitually take. Claiming, in the same studies, at least parity with anyone that exists, these men see that the conclusions they have been putting forth as true are later exposed by someone else, and shown to be false by short and easy reasoning. I shall not call their reaction envy, which then usually transforms itself into rage and hatred against those who reveal such fallacies, but I do say that they are goaded by a desire to maintain inveterate errors rather than to permit newly discovered truths to be accepted. This desire sometimes induces them to write in contradiction to those truths of which they themselves are only too aware in their own hearts, merely to keep down the reputations of other men in the estimation of the common herd of little understanding. I have heard from our Academician not a few such false conclusions, accepted as true and [yet] easy to refute; and I have kept a record of some of these.

Sagr. And you must not keep them from us, but must share them with us some time, even if we need a special session for the purpose. But now, taking up our thread again, it seems to me that we have at this point fixed the definition of uniformly accelerated motion, of which we shall treat in the ensuing discussion; and it is this:

[DEFINITION]

We shall call that motion equably or uniformly accelerated which, abandoning rest, adds on to itself equal momenta of swiftness in equal times.

Salv. This definition established, the Author requires and takes as true one single assumption; that is:

[POSTULATE]

I assume that the degrees of speed acquired by the same moveable over different inclinations of planes are equal whenever the heights of those planes are equal.

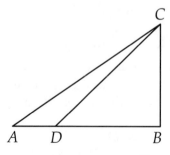

He calls the "height" of an inclined plane that vertical from the upper end of the plane which falls on the horizontal line extended through the lower end of the said inclined plane. For an understanding of this, take line AB parallel to the horizon, upon which are the two inclined planes CA and CD; the vertical CB, falling to the horizontal BA, is called by the Author the height [or altitude, or elevation] of planes CA and CD. Here he assumes that the degrees of speed of the same moveable, descending along the inclined planes CA and CD to points A and D, are equal, because their height is the same CB; and the like is also to be understood of the degree of speed that the same body falling from the point C would have at B.

Sagr. This assumption truly seems to me to be so probable as to be granted without argument, supposing always that all accidental and external impediments are removed, and that the planes are quite solid and smooth, and that the moveable is of perfectly round shape, so that both plane and moveable alike have no roughness. With all obstacles and impediments removed, my good sense [*il lume naturale*] tells me without difficulty that a heavy and perfectly round ball, descending along the lines CA, CD, and CB, would arrive at the terminal points A, D, and B with equal impetus.

Salv. You reason from good probability. But apart from mere plausibility, I wish to increase the probability so much by an experiment that it will fall little short of equality with necessary demonstration. Imagine this page to be a vertical wall, and that from a nail driven into it, a lead ball of one or two ounces hangs vertically, suspended by a fine thread two or three braccia in length, AB. Draw on the wall a horizontal line DC, cutting at right angles the vertical AB, which hangs a couple of inches out from the wall; then, moving the thread AB with its ball to AC, set the ball free. It will be seen first to descend, describing the arc CB, and then to pass the point B, running along the arc BD and rising almost up to the parallel marked CD, falling short of this by a very small interval and being prevented from arriving there exactly by the impediment of the air and the thread. From this we can truthfully conclude that the impetus acquired by the ball at point B in descent through arc CB was sufficient to drive it back up again to the same height through a similar arc BD. Having made and repeated this experiment several times, let us fix in the fall along the vertical AB, as at E or F, a nail extending out several inches, so that the thread AC, moving as before to carry the ball C through the arc CB, is stopped when it comes to

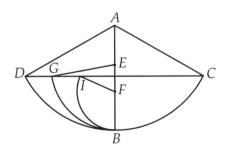

B by this nail, E, and is constrained to travel along the circumference BG, described about the center E. We shall see from this that the same

impetus can be made that, when reached at B before, drove this same moveable through the arc BD to the height of horizontal CD, but now, gentlemen, you will be pleased to see that the ball is conducted to the horizontal at point G. And the same thing happens if the nail is placed lower down, as at F, whence the ball will describe the arc BI, ending its rise always precisely at the same line, CD. If the interfering nail is so low that the thread advancing under it could not get up to the height CD, as would happen when the nail was closer to point B than to the intersection of AB with the horizontal CD, then the thread will ride on the nail and wind itself around it.

This experiment leaves no room for doubt as to the truth of our assumption, for the two arcs CB and DB being equal and similarly situated, the acquisition of momentum made by descent through the arc CB is the same as that made by descent through the arc DB; but the momentum acquired at B through arc CB is able to drive the same moveable back up through arc BD, whence also the momentum acquired in the descent DB is equal to that which drives the same moveable through the same arc from B to D. So that in general, every momentum acquired by descent through an arc equals one which can make the same moveable rise through that same arc; and all the momenta that make it rise through all the arcs BD, BG, and BI are equal, because they are created by the same momentum acquired through the descent CB, as experiment shows. Hence all the momenta acquired through descents along arcs DB, GB, and IB are equal.

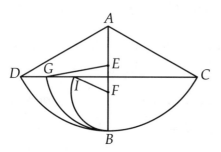

Sagr. The argument appears to me conclusive, and the experiment is so well adapted to verify the postulate that it may very well be worthy of being conceded as if it had been proved.

Salv. I do not want any of us to assume more than need be, Sagredo; especially because we are going to make use of this assumption chiefly in motions made along straight surfaces, and not curved ones, in which acceleration proceeds by degrees very different from those that we assume it to take when it proceeds in straight lines. The experiment adduced thus shows us that descent through arc CB confers such momentum on the moveable as to reconduct it to the same height along any of the arcs BD, BG, or BI. But we cannot show on this evidence that the same would happen when [even] a most perfect sphere is to descend along straight planes inclined according to the tilt of the chords of those arcs.

Indeed, we may believe that since straight planes would form angles at point *B*, a ball that had descended along the incline through the chord *CB* would encounter obstruction from planes ascending according to chords *BD*, *BG*, or *BI*; and in striking against those, it would lose some of its impetus, so that in rising it could not get back to the height of line *CD*. But if the obstacle that prejudices this experiment were removed, it seems to me that the mind understands that the impetus, which in fact takes [its] strength from the amount of the drop, would be able to carry the moveable back up to the same height.

Hence let us take this for the present as a postulate, of which the absolute truth will be later established for us by our seeing that other conclusions, built on this hypothesis, do indeed correspond with and exactly conform to experience.

This postulate alone having been assumed by the Author, he passes on to the propositions, providing them demonstratively; and the first is this:

PROPOSITION I. THEOREM I

The time in which a certain space is traversed by a moveable in uniformly accelerated movement from rest is equal to the time in which the same space would be traversed by the same moveable carried in uniform motion whose degree of speed is one-half the maximum and final degree of speed of the previous, uniformly accelerated, motion.

Let line AB represent the time in which the space CD is traversed by a moveable in uniformly accelerated movement from rest at C. Let EB, drawn in any way upon AB, represent the maximum and final degree of speed increased in the instants of the time AB. All the lines reaching AE from single points of the line AB and drawn parallel to BE will represent the increasing degrees of speed after the instant A. Next, I bisect BE at F, and I draw FG and AG parallel to BA and BF; the parallelogram AGFB will [thus] be constructed, equal to the triangle AEB, its side GF bisecting AE at I.

Now if the parallels in triangle AEB are extended as far as IG, we shall have the aggregate of all parallels contained in the quadrilateral equal to the aggregate of those included in triangle AEB, for those in triangle IEF are matched by those contained in triangle GIA, while those which are in the trapezium

AIFB are common. Since each instant and all instants of time AB corre-
spond to each point and all points of line AB, from which points the parallels
drawn and included within triangle AEB represent increasing degrees of the
increased speed, while the parallels contained within the parallelogram repre-
sent in the same way just as many degrees of speed not increased but equable,
it appears that there are just as many momenta of speed consumed in the ac-
celerated motion according to the increasing parallels of triangle AEB, as in
the equable motion according to the parallels of the parallelogram GB. For
the deficit of momenta in the first half of the accelerated motion (the momenta
represented by the parallels in triangle AGI falling short) is made up by the
momenta represented by the parallels of triangle IEF.

It is therefore evident that equal spaces will be run through in the same
time by two moveables, of which one is moved with a motion uniformly accel-
erated from rest, and the other with equable motion having a momentum
one-half the momentum of the maximum speed of the accelerated motion;
which was [the proposition] intended.

PROPOSITION II. THEOREM II

If a moveable descends from rest in uniformly accelerated motion, the
spaces run through in any times whatever are to each other as the duplicate
ratio of their times; that is, are as the squares of those times.

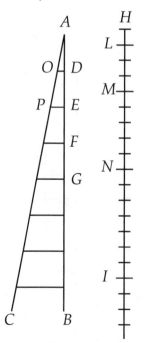

Let the flow of time from some first instant A
be represented by the line AB, in which let there
be taken any two times, AD and AE. Let HI be
the line in which the uniformly accelerated move-
able descends from point H as the first beginning
of motion; let space HL be run through in the first
time AD, and HM be the space through which it
descends in time AE. I say that space MH is
to space HL in the duplicate ratio of time EA to
time AD. Or let us say that spaces MH and HL
have the same ratio as do the squares of EA and
AD.

Draw line AC at any angle with AB. From
points D and E draw the parallels DO and EP, of
which DO will represent the maximum degree of
speed acquired at instant D of time AD, and PE
the maximum degree of speed acquired at instant
E of time AE. Since it was demonstrated above
that as to spaces run through, those are equal to

one another of which one is traversed by a moveable in uniformly accelerated motion from rest, and the other is traversed in the same time by a moveable carried in equable motion whose speed is one-half the maximum acquired in the accelerated motion, it follows that spaces MH and LH are the same that would be traversed in times EA and DA in equable motions whose speeds are as the halves of PE and OD. Therefore if it is shown that these spaces MH and LH are in the duplicate ratio of the times EA and DA, what is intended will be proved.

Now in Proposition IV of Book I ["On Uniform Motion," above] it was demonstrated that the spaces run through by moveables carried in equable motion have to one another the ratio compounded from the ratio of speeds and from the ratio of times. Here, indeed, the ratio of speeds is the same as the ratio of times, since the ratio of one-half PE to one-half OD, or of PE to OD, is that of AE to AD. Hence the ratio of spaces run through is the duplicate ratio of the times; which was to be demonstrated.

It also follows from this that this same ratio of spaces is the duplicate ratio of the maximum degrees of speed; that is, of lines PE and OD, since PE is to OD as EA is to DA.

COROLLARY I

From this it is manifest that if there are any number of equal times taken successively from the first instant or beginning of motion, say AD, DE, EF, and FG, in which spaces HL, LM, MN, and NI are traversed, then these spaces will be to one another as are the odd numbers from unity, that is, as 1, 3, 5, 7; but this is the rule [ratio] for excesses of squares of lines equally exceeding one another [and] whose [common] excess is equal to the least of the same lines, or, let us say, of the squares successively from unity. Thus when the degrees of speed are increased in equal times according to the simple series of natural numbers, the spaces run through in the same times undergo increases according with the series of odd numbers from unity.

Sagr. Please suspend the reading for a bit, while I develop a fancy that has come to my mind about a certain conception. To explain this, and for my own as well as for your clearer understanding. I'll draw a little diagram. I imagine by this line *AI* the progress of time after the first instant at *A*; and going from *A* at any angle you wish, I draw the straight line *AF*. And joining points *I* and *F*, I divide the time *AI* at the middle in *C*, and I draw *CB* parallel to *IF*, taking *CB* to be the maximum degree of the speed which, commencing from rest at *A*, grows according to the increase of the parallels to *BC* extended in triangle *ABC*; which is the same as to increase [according] as the time increases.

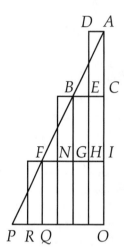

I assume without argument, from the discussion up to this point, that the space passed by the moveable falling with its speed increased in the said way is equal to the space that would be passed by the same moveable if it were moved during the same *AC* in uniform motion whose degree of speed was equal to *EC*, one-half of *BC*. I now go on to imagine the moveable [to have] descended with accelerated motion and to be found at instant *C* to have the degree of speed *BC*. It is manifest that if it continued to be moved with the same degree of speed *BC*, without accelerating further, then in the ensuing time *CI* it would pass a space double that which it passed in the equal time *AC* with degree of uniform speed *EC*, one-half the degree *BC*. But since the moveable descends with speed always uniformly increased in all equal times, it will add to the degrees *CB*, in the ensuing time *CI*, those same momenta of speed growing according to the parallels of triangle *BFG*, equal to triangle *ABC*; so that to the degree of speed *GI* there being added one-half the degree *FG*, the maximum of those [speeds] acquired in the accelerated motion governed by the parallels of triangle *BFG*, we shall have the degree of speed *IN*, with which it would be moved with uniform motion during time *CI*. That degree *IN* is triple the degree *EC* convinces [us] that the space passed in the second time *CI* must be triple that [which was] passed in the first time *CA*.

And if we assume added to *AI* a further equal part of time *IO*, and enlarge the triangle out to *APO*, then it is manifest that if the motion continued through the whole time *IO* with the degree of speed *IF* acquired in the accelerated motion during time *AI*, this degree *IF* being quadruple *EC*, the space passed in time *IO* would be quadruple that passed in the first equal time *AC*. Continuing the growth of uniform acceleration in triangle *FPQ*, similar to that of triangle *ABC* which, reduced to equable motion, adds the degree equal to *EC*, and adding *QR* equal to *EC*, we shall have the entire equable speed exercised over time *IO* quintuple the equable [speed] of the first time *AC*; and hence the space passed [will be] quintuple that [which was] passed in the first time *AC*.

Thus you see also, in this simple calculation, that the spaces passed in equal times by a moveable which, parting from rest, acquires speed in agreement with the growth of time, are to one another as the odd numbers from unity, 1, 3, 5; and taking jointly the spaces passed, that which is passed in double the time is four times that passed in the half [i.e., in the given time], and that passed in triple the time is nine times [as great]. And in short, the spaces passed are in the duplicate ratio of the times; that is, are as the squares of those times.

Simp. Really I have taken more pleasure from this simple and clear reasoning of Sagredo's than from the (for me) more obscure demonstration of the Author, so that I am better able to see why the matter must proceed in this way, once the definition of uniformly accelerated motion has been postulated and accepted. But I am still doubtful whether this is the acceleration employed by nature in the motion of her falling heavy bodies. Hence, for my understanding and for that of other people like me, I think that it would be suitable at this place [for you] to adduce some experiment from those (of which you have said that there are many) that agree in various cases with the demonstrated conclusions.

Salv. Like a true scientist, you make a very reasonable demand, for this is usual and necessary in those sciences which apply mathematical demonstrations to physical conclusions, as may be seen among writers on optics, astronomers, mechanics, musicians, and others who confirm their principles with sensory experiences that are the foundations of all the resulting structure. I do not want to have it appear a waste of time [*superfluo*] on our part, [as] if we had reasoned at excessive length about this first and chief foundation upon which rests an immense framework of infinitely many conclusions—of which we have only a tiny part put down in this book by the Author, who will have gone far to open the entrance and portal that has until now been closed to speculative minds. Therefore as to the experiments: the Author has not failed to make them, and in order to be assured that the acceleration of heavy bodies falling naturally does follow the ratio expounded above, I have often made the test [*prova*] in the following manner, and in his company.

In a wooden beam or rafter about twelve braccia long, half a braccio wide, and three inches thick, a channel was rabbeted in along the narrowest dimension, a little over an inch wide and made very straight; so that this would be clean and smooth, there was glued within it a piece of vellum, as much smoothed and cleaned as possible. In this there was made to descend a very hard bronze ball, well rounded and polished, the beam having been tilted by elevating one end of it above the horizontal plane from one to two braccia, at will. As I said, the ball was allowed to descend along [*per*] the said groove, and we noted (in the manner I shall presently tell you) the time that it consumed in running all the way, repeating the same process many times, in order to be quite sure as to the amount of time, in which we never found a difference of even the tenth part of a pulse-beat.

This operation being precisely established, we made the same ball descend only one-quarter the length of this channel, and the time of its descent being measured, this was found always to be precisely one-half the other. Next making the experiment for other lengths, examining now the time for the whole length [in comparison] with the time of one-half, or with that of two-thirds, or of three-quarters, and finally with any other division, by experiments repeated a full hundred times, the spaces were always found to be to one another as the squares of the times. And this [held] for all inclinations of the plane; that is, of the channel in which the ball was made to descend, where we observed also that the times of descent for diverse inclinations maintained among themselves accurately that ratio that we shall find later assigned and demonstrated by our Author.

As to the measure of time, we had a large pail filled with water and fastened from above, which had a slender tube affixed to its bottom, through which a narrow thread of water ran; this was received in a little beaker during the entire time that the ball descended along the channel or parts of it. The little amounts of water collected in this way were weighed from time to time on a delicate balance, the differences and ratios of the weights giving us the differences and ratios of the weights giving us the differences and ratios of the times, and with such precision that, as I have said, these operations repeated time and again never differed by any notable amount.

Simp. It would have given me great satisfaction to have been present at these experiments. But being certain of your diligence in making them and your fidelity in relating them, I am content to assume them as most certain and true.

Salv. Then we may resume our reading, and proceed.

Commentary on *Two New Sciences*, pp. 153–170

One of Galileo's basic arguments is that bodies in free fall move in such a way that their speed at any point is proportional to their times of fall, the proportionality constant being a, the acceleration, which Galileo claims is constant. His result can be encapsulated in the equation $V_f = at$. About two-thirds of the way through this selection, Galileo derives two theorems that apply to cases in which there is constant acceleration. You should recognize the first theorem because you have seen it before. It is essentially the Merton Mean Speed Theorem, which was discussed previously. It had been published by various sixteenth-century authors in whose writings

Galileo may have seen it. Whatever the case, it is one of the cornerstones of Galilean mechanics. We can formulate the content of Galileo's first theorem in symbolic form by means of the equation that states, acceleration being constant,

$$\frac{V_o + V_f}{2} \cdot t = s,$$

where V_o is the initial speed of a body experiencing a constant acceleration, V_f is its final speed, t is the time during which the body is in motion, and s is the distance through which it moves. It should be noted that the above equation somewhat extends the meaning of Galileo's first theorem, which does not explicitly discuss initial velocities with non-zero values. Galileo's second theorem can be stated as

$$\frac{s_1}{s_2} = \frac{t_1^2}{t_2^2},$$

which put in verbal form states that the distances traversed by two bodies experiencing constant accelerations will be as the squares of their times of motion. Galileo's derivation of this formula is quite simple when put into symbolic form. Substituting the formula $V_f = V_0 + at$, which he had proved earlier in his book, into his first equation, we have

$$s = \frac{V_o + V_f}{2} \cdot t = \frac{V_o + V_o + a \cdot t}{2} \cdot t = V_o t + \frac{at^2}{2}.$$

For the case where the initial speed is 0, we have

$$s = \frac{at^2}{2}.$$

This is highly useful equation for dealing with problems of motion.

As to Galileo's method, the statement by Salviati that begins "The present does not seem to me to be an opportune time to enter into the investigation of the cause of the acceleration of natural motion..."[13] is especially significant. You should see from it that Galileo's concern, at least at this stage, was primarily **kinematic** rather than dynamic. He is concerned with **describing how** bodies move rather than with **explaining why** they move as they do. In this sense, he should be seen not as having answered all the questions about motion that Aristotle had discussed, but rather as having selected from those questions a subset of questions that he believed he could answer. It is especially important to note that Galileo is setting aside the question of the nature of gravity.

---

[13] Galileo, *Two New Sciences,* pp. 158–159.

The prior existence of the Merton Mean Speed Theorem may seem to raise questions about the degree of originality to be ascribed to Galileo. The mathematics of uniformly accelerated motion had after all already been worked out. This raised the question: What constitutes the originality in Galileo's contribution? The answer is in Galileo proving, as Simplicio demands that he do, that bodies in free fall actually conform to uniform acceleration. As noted before, this could not be done directly; one could not simply observe that free fall occurs in a uniformly accelerated manner. In fact, Galileo never in his *Two New Sciences* determined that bodies in free fall accelerate at 32 feet per second per second. This constant was first determined by Christiaan Huygens later in the century.

The main method used by Galileo in proving that bodies in free fall accelerate uniformly is based on the inclined plane. This device slows down the motion of bodies to the point that their motion can be directly observed. Unfortunately, Galileo recounts his inclined plane experiments without providing quantified data. Attempts over the last few decades to duplicate Galileo's results have, however, been moderately successful. The difficulties in this should be fairly clear. One needs a precise method of measuring small intervals of time. Friction must be reduced as far as possible. The determinations of starting time and finishing time must be very accurate. Moreover, questions can be raised about whether such experiments can ever be conclusive. Galileo's argument is that if one finds that the $s \propto t^2$ relation holds for ever more steeply inclined planes, then the relationship should hold for the case of a "vertically" inclined plane, that is, for a body in free fall. How satisfactory do you find this argument?

Most of the remainder of the "Third Day" of Galileo's *Two New Sciences* continues the geometrical discussion of motion on inclined planes. Typical of the theorems developed is Galileo's Proposition V, Theorem V,[14] which proposition states that for t (time of fall), h (height of plane), and L (length of plane),

$$\frac{t_1}{t_2} = \frac{L_1}{L_2} \cdot \frac{\sqrt{h_2}}{\sqrt{h_1}}.$$

We need not follow this discussion further.

Problems

4. A five pound brass ball is dropped from a height of 400 feet. Calculate (1) its time of fall, (2) its final speed, and (3) its average speed. In working this problem, use 32 feet per second^2 as the acceleration.
Answers: (1) 5 seconds. (2) 160 feet per second. (3) 80 feet per second.

---

[14]Galileo, *Two New Sciences*, p. 177.

5. Upon dropping a piano from the top of a tall building, you notice that it takes 7 seconds to reach the ground. Determine the height of the building. Answer: 784 feet.

6. An eight pound iron ball slides down a frictionless inclined plane, the base angle of which is 47° and the height of which is sixteen feet. Calculate the final speed of the ball. Explain your method in making this calculation. Answer: 32 feet per second.

7. For a certain inclined plane, the rate of acceleration is equal to 2 feet per second^2. Determine the position and speed of a ball rolling down the plane for each of the times indicated in the table below. Also determine the distance traveled during each second. Then reread Corollary I to Theorem II in the selection from Galileo's book.

time	distance covered at the end of this time	speed attained	distance covered during this second
1 sec			
2 sec			
3 sec			
4 sec			
5 sec			

Galileo, the Law of Inertia, and Projectile Motion

To attain an understanding of Galileo's ideas on projectile motion, it is best to begin with a discussion of the development of the **law of inertia**. The history of the idea of inertia extends from Aristotle to Newton and even to Einstein and is filled with fascinating twists and paradoxes. Moreover, at each stage, ideas of inertia were crucial for comprehension of how bodies move.

Aristotle was convinced that the continued motion of a projectile requires explanation; why should a ball persist in motion even after leaving the hand of the person throwing it? Such motion, he reasoned, is not natural but rather "violent" motion and as such demands explanation. Aristotle's position as stated in his *Physics* is that

> things which are thrown are in motion, though that which pushed them is not touching them, either because of mutual replacement, as some say, or because the air that has been pushed pushes [them] with a motion which is faster than the locomotion of the things pushed, that is, the locomotion with which those things travel to their *proper* place.[15]

---

[15] *Aristotle's Physics*, trans. by Hippocrates G. Apostle (Peripatetic Press, 1980), p. 73. See Aristotle, *Physics*, Bk. IV, Ch. 8, 215a 14.

One curious feature of Aristotle's position is that according to it the medium through which the body moves acts both to retard *and* to perpetuate the motion.

A number of authors criticized this Aristotelian doctrine; for example, **Johannes Philoponus**, an early sixth century Christian neo-Platonist, in his *Commentary on Aristotle's Physics* pointed out problems with this doctrine, suggesting in opposition to it that *"some incorporeal motive force is imparted by the projector to the projectile. . . ."*[16] The Arabic scholar **Avicenna** (d. 1037) also objected to Aristotle's position. Avicenna urged that projectiles have impressed upon them a force known as *mail* (meaning inclination or tendency), which being a permanent quality would keep them in motion eternally, were the object to encounter no resistance. Bodies are capable of receiving *mail* in proportion to their weight. Another Arab, **Abul Barakat** (d. ca.1164), propounded a somewhat different version of the *mail* theory, urging that *mail* is self-dissipating and that consequently even in a vacuum bodies would finally cease their motion.

Passing over a number of other authors who discussed this topic, we come to the Parisian scholar **Jean Buridan** (ca.1300–ca.1358), who in his *Questions on the Eight Books of the* Physics [of Aristotle] presented a particularly important form of this doctrine. Buridan argued against Aristotle's idea that the medium rushes into space behind the thrown projectile, pushing it forward. He noted that tops and smith wheels, for example, continue their motion even though the medium cannot account for this. Also javelins, pointed at *both* ends, carry through long distances. Buridan's own position, which may have had Arabic roots, was that an **impetus** is impressed on the thrown body and acts to keep the body in motion against the resistance of the medium. This impetus is measured by both the weight and speed of the body. Like Avicenna, he viewed this entity as conserved and consequently believed that in the absence of resistance, a body would move eternally. He even speculated that God in creating the universe could have bestowed on each planet the specific quantity of impetus necessary to keep it in continued motion through the resistance-free heavens. This, as he noted, would explain their motions without the necessity of attributing intellects or angels to them as the source of their motions. Buridan also applied his impetus doctrine to the accelerating fall of bodies. His idea was that bodies in falling are at first moved only by their tendency to move to their natural place, but that with the speed thus attained, they acquire an impetus, which gradually accumulates, making the body at each stage move faster.

---

[16]As given in Morris R. Cohen and I. E. Drabkin (eds.), *A Source Book in Greek Science* (Harvard University Press, 1958), p. 223.

Two pre-Galilean views on projectile motion: the picture at the left was published in 1547 by Gualtherius Rivius, that at the right in 1561 by Daniele Santbech.

In the period between Buridan and Galileo, a number of authors grappled with these issues, yet in Galileo's early *De Motu,* he advocated a doctrine of impetus very similar to Buridan's, except that Galileo's impetus was self-dissipating. Moreover, it should be clear that in a number of important ways the impetus theory helped Galileo toward the form of the principle of inertia that he eventually espoused. In seeing this point and also in recognizing the significance of this idea for the Copernican debate, you will be helped by reading the following pages from Galileo's *Dialogue Concerning the Two Chief World Systems,* which, as noted previously, was published in 1632, six years before his *Two New Sciences.*

In the sections immediately before these materials, Galileo discussed the objection to the Copernican system that if the Earth moves, a body dropped from a tower ought to fall some distance from the tower just as a body dropped from the mast of a moving ship ought to fall toward the stern of the ship. Opponents of Copernicanism also state that cannon balls fired directly up ought to fall at some distance from the point from which they were fired and that similarly bullets fired to the east or west, north or south, ought to have their paths influenced by the motion of the Earth. In this context, special attention is given to the problem of the path of weights dropped from a tower on a moving Earth. In this case, the Aristotelian argument is that bodies on a stationary Earth should fall straight to the ground. Thus they urged that because observation shows that bodies do fall straight to the ground provides strong evidence that the Earth is stationary. In response to this, Galileo suggested that it could be that a body dropped from a tower on a moving Earth will also fall straight to the ground. If this is the case, then the Aristotelian argument fails. Aristotelians objected moreover that were it the case that a body on a moving Earth would fall directly to the ground, the body would have to have not one but two motions, a notion that they found unacceptable. Galileo will later show that this is not only possible, but that the motions involved are compounded in such a manner that straight line motion results.

With this as background let us turn to Galileo's text.

Galileo, *Two Chief World Systems*, pp. 143–149[17]

Simp. So far, yes; and though you have brought up some trivial disparities, they do not seem to me of such moment as to suffice to shake my conviction.

Salv. Rather, I hope that you will stick to it, and firmly insist that the result on the earth must correspond to that on the ship, so that when the latter is perceived to be prejudicial to your case you will not be tempted to change your mind. You say, then, that since when the ship stands still the rock falls to the foot of the mast, and when the ship is in motion it falls apart from there, then conversely, from the falling of the rock at the foot it is inferred that the ship stands still, and from its falling away it may deduced that the ship is moving. And since what happens on the ship must likewise happen on the land, from the falling of the rock at the foot of the tower one necessarily infers the immobility of the terrestrial globe. Is that your argument?

Simp. That is exactly it, briefly stated, which makes it easy to understand.

Salv. Now tell me: If the stone dropped from the top of the mast when the ship was sailing rapidly fell in exactly the same place on the ship to which it fell when the ship was standing still, what use could you make of this falling with regard to determining whether the vessel stood still or moved?

Simp. Absolutely none; just as by the beating of the pulse, for instance, you cannot know whether a person is asleep or awake, since the pulse beats in the same manner in sleeping as in waking.

Salv. Very good. Now, have you ever made this experiment of the ship?

Simp. I have never made it, but I certainly believe that the authorities who adduced it had carefully observed it. Besides, the cause of the difference is so exactly known that there is no room for doubt.

Salv. You yourself are sufficient evidence that those authorities may have offered it without having performed it, for you take it as

---

[17]Galileo Galilei, *Dialogue Concerning the Two Chief World Systems—Ptolemaic and Copernican*, trans. by Stillman Drake (University of California Press, 1953).

certain without having done it, and commit yourself to the good faith of their dictum. Similarly it not only may be, but must be that they did the same thing too—I mean, put faith in their predecessors, right on back without ever arriving at anyone who had performed it. For anyone who does will find that the experiment shows exactly the opposite of what is written; that is, it will show that the stone always falls in the same place on the ship, whether the ship is standing still or moving with any speed you please. Therefore, the same cause holding good on the earth as on the ship, nothing can be inferred about the earth's motion or rest from the stone falling always perpendicularly to the foot of the tower.

Simp. If you had referred me to any other agency than experiment, I think that our dispute would not soon come to an end; for this appears to me to be a thing so remote from human reason that there is no place in it for credulity or probability.

Salv. For me there is, just the same.

Simp. So you have not made a hundred tests, or even one? And yet you freely declare it to be certain? I shall retain my credulity, and my own confidence that the experiment has been made by the most important authors who make use of it, and that it shows what they say it does.

Salv. Without experiment I am sure that the effect will happen as I tell you, because it must happen that way; and I might add that you yourself also know that it cannot happen otherwise, no matter how you may pretend not to know it—or give the impression. But I am so handy at picking people's brains that I shall make you confess this in spite of yourself.

Sagredo is very quiet; it seemed to me that I saw him move as though he were about to say something.

Sagr. I was about to say something or other, but the interest aroused in me by hearing you threaten Simplicio with this sort of violence in order to reveal the knowledge he is trying to hide has deprived me of any other desire; I beg you to make good your boast.

Salv. If only Simplicio is willing to reply to my interrogation, I cannot fail.

Simp. I shall reply as best I can, certain that I shall be put to little trouble; for of the things I hold to be false, I believe I can know nothing, seeing that knowledge is of the true and not of the false.

Salv. I do not want you to declare or reply anything that you do not know for certain. Now tell me: Suppose you have a plane surface as smooth as a mirror and made of some hard material like steel. This is not parallel to the horizon, but somewhat inclined, and upon it you have placed a ball which is perfectly spherical and of some hard and heavy material like bronze. What do you believe this will do when released? Do you not think, as I do, that it will remain still?

Simp. If that surface is tilted?

Salv. Yes, that is what was assumed.

Simp. I do not believe that it would stay still at all; rather, I am sure that it would spontaneously roll down.

Salv. Pay careful attention to what you are saying, Simplicio, for I am certain that it would stay wherever you placed it.

Simp. Well, Salviati, so long as you make use of assumptions of this sort I shall cease to be surprised that you deduce such false conclusions.

Salv. Then you are quite sure that it would spontaneously move downward?

Simp. What doubt is there about this?

Salv. And you take this for granted not because I have taught it to you—indeed, I have tried to persuade you to the contrary—but all by yourself, by means of your own common sense.

Simp. Oh, now I see your trick; you spoke as you did in order to get me out on a limb, as the common people say, and not because you really believed what you said.

Salv. That was it. Now how long would the ball continue to roll, and how fast? Remember that I said a perfectly round ball and a highly polished surface, in order to remove all external and accidental impediments. Similarly I want you to take away any impediment of the air caused by its resistance to separation, and all other accidental obstacles, if there are any.

Simp. I completely understood you, and to your question I reply that the ball would continue to move indefinitely, as far as the slope of the surface extended, and with a continually accelerated motion. For such is the nature of heavy bodies, which *vires acquirunt eundo*;[18] and the greater the slope, the greater would be the velocity.

Salv. But if one wanted the ball to move upward on this same surface, do you think it would go?

Simp. Not spontaneously, no; but drawn or thrown forcibly, it would.

Salv. And if it were thrust along with some impetus impressed forcibly upon it, what would its motion be, and how great?

---

[18]This can be translated as "Gain strength as they go." It is a reference to a passage in Virgil's *Aeneid*, iv, 175.

Simp. The motion would constantly slow down and be retarded, being contrary to nature, and would be of longer or shorter duration according to the greater or lesser impulse and the lesser or greater slope upward.

Salv. Very well; up to this point you have explained to me the events of motion upon two different planes. On the downward inclined plane, the heavy moving body spontaneously descends and continually accelerates, and to keep it at rest requires the use of force. On the upward slope, force is needed to thrust it along or even to hold it still, and motion which is impressed upon it continually diminishes until it is entirely annihilated. You say also that a difference in the two instances arises from the greater or lesser upward or downward slope of the plane, so that from a greater slope downward there follows a greater speed, while on the contrary upon the upward slope a given movable body thrown with a given force moves farther according as the slope is less.

Now tell me what would happen to the same movable body placed upon a surface with no slope upward or downward.

Simp. Here I must think a moment about my reply. There being no downward slope, there can be no natural tendency toward motion; and there being no upward slope, there can be no resistance to being moved, so there would be an indifference between the propensity and the resistance to motion. Therefore it seems to me that it ought naturally to remain stable. But I forgot; it was not so very long ago that Sagredo gave me to understand that that is what would happen.

Salv. I believe it would do so if one set the ball down firmly. But what would happen if it were given an impetus in any direction?

Simp. It must follow that it would move in that direction.

Salv. But with what sort of movement? One continually accelerated, as on the downward plane, or increasingly retarded as on the upward one?

Simp. I cannot see any cause for acceleration or deceleration, there being no slope upward or downward.

Salv. Exactly so. But if there is no cause for the ball's retardation, there ought to be still less for its coming to rest; so how far would you have the ball continue to move?

Simp. As far as the extension of the surface continued without rising or falling.

Salv. Then if such a space were unbounded, the motion on it would likewise be boundless? That is, perpetual?

Simp. It seems so to me, if the movable body were of durable material.

Salv. That is of course assumed, since we said that all external and accidental impediments were to be removed, and any fragility on the part of the moving body would in this case be one of the accidental impediments.

Now tell me, what do you consider to be the cause of the ball moving spontaneously on the downward inclined plane, but only by force on the one tilted upward?

Simp. That the tendency of heavy bodies is to move toward the center of the earth, and to move upward from its circumference only with force; now the downward surface is that which gets closer to the center, while the upward one gets farther away.

Salv. Then in order for a surface to be neither downward nor upward, all its parts must be equally distant from the center. Are there any such surfaces in the world?

Simp. Plenty of them; such would be the surface of our terrestrial globe if it were smooth, and not rough and mountainous as it is. But there is that of the water, when it is placid and tranquil.

Salv. Then a ship, when it moves over a calm sea, is one of these movables which courses over a surface that is tilted neither up nor down, and if all external and accidental obstacles were removed, it would thus be disposed to move incessantly and uniformly from an impulse once received?

Simp. It seems that it ought to be.

Salv. Now as to that stone which is on top of the mast; does it not move, carried by the ship, both of them going along the circumference of a circle about its center? And consequently is there not in it an ineradicable motion, all external impediments being removed? And is not this motion as fast as that of the ship?

Simp. All this is true, but what next?

Salv. Go on and draw the final consequence by yourself, if by yourself you have known all the premises.

Simp. By the final conclusion you mean that the stone, moving with an indelibly impressed motion, is not going to leave the ship, but will follow it, and finally will fall at the same place where it fell when the ship remained motionless. And I, too, say that this would follow if there were no external impediments to disturb the motion of the stone after it was set free. But there are two such impediments; one is the inability of the movable body to split the air with its own impetus alone, once it has lost the force from the oars which it shared as part of the ship while it was on the mast; the other is the new motion of falling downward, which must impede its other, forward, motion.

Salv. As for the impediment of the air, I do not deny that to you, and if the falling body were of very light material, like a feather or a tuft of wool, the retardation would be quite considerable. But in a heavy stone it is insignificant, and if, as you yourself just said a little while ago, the force of the wildest wind is not enough to move a large stone from its place, just imagine how much the quiet air could accomplish upon meeting a rock which moved no faster than the ship! All the same, as I said, I concede to you the small effect which may depend upon such an impediment, just as I know you will concede to me that if the air were moving at the same speed as the ship and the rock, this impediment would be absolutely nil.

As for the other, the supervening motion downward, in the first place it is obvious that these two motions (I mean the circular around the center and the straight motion toward the center) are not contraries, nor are they destructive of one another, nor incompatible. As to the moving body, it has no resistance whatever to such a motion, for you yourself have already granted the resistance to be against motion which increases the distance from the center, and the tendency to be toward motion which approaches the center. From this it follows necessarily that the moving body has neither a resistance nor a propensity to motion which does not approach toward or depart from the center, and in consequence no cause for diminution in the property impressed upon it. Hence the cause of motion is not a single one which must be weakened by the new action, but there exist two distinct causes. Of these, heaviness attends only to the drawing of the movable body toward the center, and impressed force only to its being led around the center, so no occasion remains for any impediment.

Later in Galileo's *Two Chief World Systems*, Salviati presents the following:

Galileo, *Two Chief World Systems*, pp. 186–188

[*Salv.*] For a final identification of the nullity of the experiments brought forth, this seems to me the place to show you a way to test them all very easily. Shut yourself up with some friend in the main cabin below decks on some large ship, and have with you there some flies, butterflies, and other small flying animals. Have a large bowl of water with some fish in it; hang up a bottle that empties drop by drop into a wide vessel beneath it. With the ship standing still, observe carefully how the little animals fly with equal speed to all sides of the cabin. The fish swim indifferently in all directions; the drops fall into the vessel beneath; and, in throwing something to your friend, you need throw it no more strongly in one direction than another, the distances being equal; jumping with your feet together, you pass equal spaces in every direction. When you have observed all these things carefully (though there is no doubt that when the ship is standing still everything must happen in this way), have the ship proceed with any speed you like, so long as the motion is uniform and not fluctuating this way and that. You will discover not the least change in all the effects named, nor could you tell from any of them whether the ship was moving or standing still. In jumping you will pass on the floor the same spaces as before, nor will you make larger jumps toward the stern than toward the prow even though the ship is moving quite rapidly, despite the fact that during the time that you are in the air the floor under you will be going in a direction opposite to your jump. In throwing something to your companion, you will need no more force to get it to him whether he is in the direction of the bow or the stern, with yourself situated opposite. The droplets will fall as before into the vessel beneath without dropping toward the stern, although while the drops are in the air the ship runs many spans. The fish in their water will swim toward the front of their bowl with no more effort than toward the back, and will go with equal ease to bait placed anywhere around the edges of the bowl. Finally the butterflies and flies will continue their flights indifferently toward every side, nor will it ever happen that they are concentrated toward the stern, as if tired out from keeping up with the course of the ship from which they will have been separated during long intervals by keeping themselves in the air. And if smoke is made by burning some incense, it will be seen going up in the form of a little cloud, remaining still and moving no more toward one side than the other. The cause of all these correspondences of effects is the fact that the ship's motion is common to all the things contained in it and to the air also. That is why I said you should be below decks; for if this took place above in the open air, which would not follow the course of the ship, more or less noticeable differences would be seen in some of the effects noted. No doubt the smoke would fall as much behind as the air itself. The flies likewise, and the butterflies, held back by the air, would be unable to follow the ship's motion if they were separated from it by a perceptible distance. But keeping themselves near it, they would follow it without effort or hindrance; for the ship, being an unbroken structure, carries with it a part of the nearby air. For a similar reason we sometimes, when riding horseback, see persistent flies and horseflies following our horses, flying now to one part of their bodies and now to another. But the difference would be small as regards the falling drops, and as to the jumping and the throwing it would be quite imperceptible.

Commentary on *Two Chief World Systems*, pp. 186–188

The most important single passage in this reading occurs when Salviati, referring to the motion of a ball on a frictionless plane, rhetorically asks: "Then if such a space were unbounded, the motion on it would likewise be boundless? That is, perpetual?" This statement is similar to Galileo's claim in his *Two New Sciences* that "... whatever degree of speed is found in the moveable, this is by its nature [*suapte natura*] indelibly impressed on it when external causes of acceleration or retardation are removed, which occurs only on the horizontal plane.... From this it... follows that motion in the horizontal is also eternal...."[19] These statements have sometimes been seen as identical to the **law of inertia**, which was later formulated by **Isaac Newton** as: "Every body continues in its state of rest, or of uniform motion in a right line, unless it is compelled to change that state by forces impressed upon it." You should, however, be clear that the statements are not identical. For example, Galileo adopted the idea of a **naturally perpetual circular motion**, comparable in some ways to the Greek idea of perfect circular motion. By means of this, he explained the planetary motions and also the fact that the Earth does not fly apart as it rotates. Secondly, the medieval idea of an impetus remained to some degree present in Galileo's analysis. On the other hand, one can see in the reading from Galileo that he has some understanding of what Einstein refers to as the **Galilean principle of relativity**. This is the idea that it is impossible to distinguish between two frames of motion moving at a constant rate relative to each other. For example, were you locked in a windowless cabin on a ship, you could not determine by dropping or throwing weights whether the ship is stationary or moving at a constant speed. Within this perspective, it is possible to argue that what must be explained in projectile motion is the *stopping* of the body, not the *continuation* of the motion. You may also be able to see within this point of view that the principle of the relativity of motion is a powerful principle in physical analysis.

If Galileo did not attain a full statement of the law of inertia, who first did so? The answer that most scholars favor is **René Descartes**, who included the principle in his *Principles of Philosophy* (1644).

---

[19] Galileo, *Two New Sciences*, trans. Drake, p. 197.

Galileo's Mathematical Treatment of Projectile Motion

The mathematics involved in Galileo's discussion of projectile motion in the "Fourth Day" of his *Two New Sciences* is somewhat complex, but the physical ideas employed are relatively simple; in fact, they are set out along with his most important conclusion in the first few paragraphs of his "Fourth Day" dialogue.

Galileo, *Two New Sciences*, p. 217

FOURTH DAY

Salv. Simplicio is just arriving now, so let us begin on motion without delay. Here is our Author's text:

On the Motion of Projectiles

We have considered properties existing in equable motion, and those in naturally accelerated motion over inclined planes of whatever slope. In the studies on which I now enter, I shall try to present certain leading essentials, and to establish them by firm demonstrations, bearing on a moveable when its motion is compounded from two movements; that is, when it is moved equably and is also naturally accelerated. Of this kind appear to be those which we speak of as projections, the origin of which I lay down as follows.

I mentally conceive of some moveable projected on a horizontal plane, all impediments being put aside. Now it is evident from what has been said elsewhere at greater length that equable motion on this plane would be perpetual if the plane were of infinite extent; but if we assume it to be ended, and [situated] on high, the moveable (which I conceive of as being endowed with heaviness), driven to the end of this plane and going on further, adds on to its previous equable and indelible motion that downward tendency which it has from its own heaviness. Thus there emerges a certain motion, compounded from equable horizontal and from naturally accelerated downward [motion], which I call "projection." We shall demonstrate some of its properties [accidentia], of which the first is this:

PROPOSITION I. THEOREM I

When a projectile is carried in motion compounded from equable horizontal and from naturally accelerated downward [motions], it describes a semiparabolic line in its movement.

Commentary on *Two New Sciences*, p. 217

The key idea in Galileo's analysis of projectile motion is that the motion of, say, a cannonball fired from a horizontal cannon atop a cliff can be understood as being compounded of two motions: (1) the constant horizontal speed given the cannonball, and (2) the vertical acceleration of the cannonball due to gravity. Galileo claims that these two motions are *independent* of each other and consequently can be conceptually separated without violating the physics of the situation. In other words, after exiting the cannon, the ball immediately begins to fall to the ground just as would another cannon ball dropped from the end of the cannon barrel at the time of firing. The first cannonball, unlike the second cannonball, has a horizontal speed carrying it in the direction in which the cannon is aimed. This horizontal speed in no way changes the time of fall of the ball; its effect is rather to cause the ball to fall to the ground at a point horizontally distant from the barrel. The next diagram represents this situation.

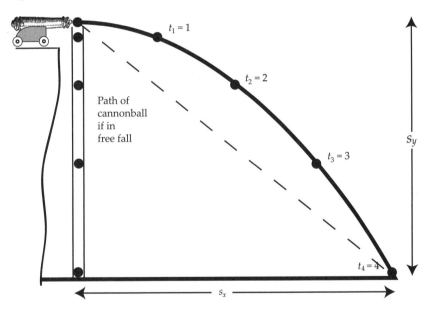

The horizontal speed v_x is the initial speed of the cannonball. After time t, the ball will have traveled a distance s_x in the horizontal direction such that $s_x = v_x t$. Simultaneously the ball will be falling to the ground at exactly the same rate as a ball dropped from the same height. The distance of fall in the vertical direction will be $s_y = \frac{g t^2}{2}$, where g is the acceleration due to gravity, that is, 32 feet per second squared. The total displacement of the fired ball will be (using the Pythagorean Theorem)

$$s = \sqrt{s_x^2 + s_y^2}.$$

Combining our s_x and s_y equations produces an interesting result. From the well known formula that $s_x = v_x t$, we can conclude that

$$t^2 = \frac{s_x{}^2}{v_x}.$$

Substituting this into the equation $s_y = \frac{gt^2}{2}$, we get

$$s_y = \frac{g(s_x/v_x)^2}{2} = \frac{gs_x{}^2}{2v_x{}^2}.$$

But because both g and v_x are fixed amounts, $\frac{g}{2v_x{}^2}$ must be a constant. Thus we have an equation of the form $s_y = ks_x{}^2$ (k being the constant). Equations of this form plot as parabolas. Consequently, the path of a projectile will be, as Galileo shows by a rather more cumbrous argument, a parabola.

Problem

8. Joe, a hungry hunter, is perched in a spot such that when he aims his rifle at a squirrel 6 feet up in a tree that is a thousand feet away, his rifle is horizontal. Joe's rifle fires bullets at a speed of 2000 feet per second. The squirrel instantaneously sees the flash of the rifle and, also instantaneously, decides to drop to the ground so as to avoid being hit. Determine what Joe had for dinner that night, showing your calculations.
(Solution to be worked out by the reader.)

Projectile Motion in General

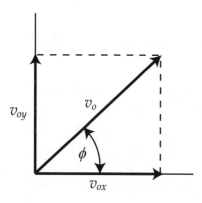

So far, the only cases considered have been those in which the projectile's initial speed was horizontal. The approach used in these cases can, however, easily be generalized. Imagine a projectile propelled at an angle ϕ (read as "phi") up from the horizontal and with an initial speed v_0. Its motion at any subsequent time can be resolved into three elements:

1. The initial speed given it in a vertical direction v_{oy}, which will equal $v_0 \sin \phi$;

2. The initial speed in the horizontal direction v_{ox}, which will equal $v_o \cos \phi$;

3. Its acceleration g toward the Earth due to gravity.

To determine its position and speed at any point, you need only combine these three elements.

Sample Problems

Problem 1. A cannon is fired at an angle $\phi = 30°$, giving the shell an initial speed of 1500 feet per sec. It is fired from the top of a fort 50 feet high. Calculate (1) when the shell will reach its highest point, (2) the time it will take during its flight, and (3) the distance it will travel.

Solution:

Part 1. This shell is fired with a vertical speed of

$$v_{oy} = \sin \phi \cdot 1500 \frac{\text{ft.}}{\text{sec.}} = .5 \cdot 1500 \frac{\text{ft.}}{\text{sec.}} = 750 \frac{\text{ft.}}{\text{sec.}}.$$

The point at which the shell reaches its greatest vertical height will be the point where its vertical speed will be zero, that is, where the downward acceleration it experiences due to gravity, has completely overcome its vertical speed upward. We can apply our formula $v_f = v_o + at$ to this case. We know that its final vertical speed, that is, its speed at its highest point, will equal 0. Thus, we have

$$0 = 750 \frac{\text{ft.}}{\text{sec.}} + \left(-32 \frac{\text{ft.}}{\text{sec.}^2}\right) \cdot t.$$

Note that we must make the acceleration negative because it is in the opposite direction from that of the vertical speed. Solving this equation for t gives

$$t = \frac{750}{32} = 23.4 \text{ sec.}$$

Part 2. By using our equation $s = \frac{at^2}{2}$, we find that the shell will then have attained a height

$$s_y = \frac{32 \frac{\text{ft.}}{\text{sec.}^2}}{2} \cdot (23.4 \text{ sec.})^2 = 16 \cdot 548 \text{ ft.} = 8,768 \text{ ft.}$$

It will have to drop this distance plus 50 feet (the height of the fort) before it hits the ground. The time necessary for this is given by the equation:

$$50 \text{ ft.} + 8768 \text{ ft.} = s = \frac{gt^2}{2} = \frac{32 \text{ ft.}/\text{sec.}^2}{2} t^2.$$

Thus we have: $t^2 = 551 \text{ sec}^2$, from which we conclude that the time of downward fall will be $t = 23.5$ sec. Consequently, the shell will be in the air a total of $(23.4 + 23.5) = 46.9$ sec.

Part 3. Its horizontal speed is

$$v_{ox} = v_o \cos \phi = 1500 \frac{\text{ft.}}{\text{sec.}} \cdot \cos 30° = 1500 \cdot .866 \frac{\text{ft.}}{\text{sec.}} = 1299 \frac{\text{ft.}}{\text{sec.}}.$$

In 46.9 seconds, it will have traveled $46.9 \text{ sec.} \cdot 1299 \frac{\text{ft.}}{\text{sec.}} = 60,923$ feet or about 11.5 miles.

Problem 2. At what angle should a shot putter aim to attain maximum distance?

Solution: First, regarding the horizontal motion, we know $s_x = v_x t = (v_o \cos \phi)t$, where ϕ represents the angle of launch measured up from the horizontal. The vertical motion of the shot is described by the equation:

$$s_y = -\frac{gt^2}{2} + v_{oy}t.^{20}$$

Since we wish to know the time at which the shot returns to the ground, let us solve for t when $s_y = 0$.

$$s_y = 0 = -\frac{gt^2}{2} + v_{oy}t,$$

which implies that

$$v_{oy}t = \frac{gt^2}{2}.$$

Because $v_{oy} = v_o \sin \phi$, we have

$$(v_o \sin \phi)t = \frac{gt^2}{2}.$$

Solving this for t gives

$$t = \frac{2v_o \sin \phi}{g}.$$

Substituting this value for t into our original equation, we get:

$$s_x = \frac{(2v_o \sin \phi)(v_o \cos \phi)}{g}.$$

---

[20] The negative sign before the gravitational constant of acceleration conveys the fact that the acceleration tends in the downward direction, whereas the vertical component v_{oy} of the initial vertical velocity will be upward.

Using the well known trigonometric equality that $2 \sin \phi \cos \phi = \sin 2\phi$, we have:

$$s_x = \frac{v_0{}^2 \sin 2\phi}{g}.$$

From an examination of this equation, we see that s_x is largest when $\sin 2\phi$ is at its maximum, which is when $\phi = 45°$. Consequently, projectiles travel farthest when they are released at a $45°$ angle.

Keeping the conclusion attained in Sample Problem 2 in mind, let us return to Galileo's *Two New Sciences* for a final short reading that illustrates a way in which Galileo applied his analysis to practical matters. This reading also provides empirical evidence for the correctness of his theory.

Galileo, *Two New Sciences*, trans. Drake, pp. 245–246

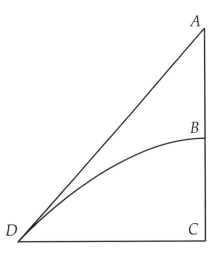

From this it is clear that in reverse [direction] through the semiparabola DB, the projectile from point D requires less impetus that through any other [semiparabola] having greater or smaller elevation than semiparabola BD, which [elevation] is according to the tangent AD and contains one-half a right angle with the horizontal. Hence it follows that if projections are made with the same impetus from point D, but according to different elevations, the maximum projection, or amplitude of semiparabola (or whole parabola) will be that corresponding to the elevation of half a right angle. The others, made according to larger or smaller angles, will be shorter [in range].

Sagr. The force of necessary demonstrations is full of marvel and delight; and such are mathematical [demonstrations] alone. I already knew, by trusting to the accounts of many bombardiers, that the maximum of all ranges of shots, for artillery pieces or mortars—that is, that shot which takes the ball farthest—is the one made at elevation of half a right angle, which they call "at the sixth point of the [gunner's] square." But to understand the reason for this phenomenon infinitely surpasses the simple idea obtained from the statements of others, or even from experience many times repeated.

Salv. You say well. The knowledge of one single effect acquired through its causes opens the mind to the understanding and certainty of other effects without need of recourse to experiments. That is exactly what happens in the present instance; for having gained by demonstrative reasoning the certainty that the maximum of all ranges of shots is that of elevation at half a right angle, the Author demonstrates to us something that has perhaps not been observed through experiment; and this is that of the other shots, those are equal [in range] to one another whose elevations exceed or fall short of half a right angle by equal angles. Thus two balls shot, one at an elevation of $52\frac{1}{2}°$ [7 *punti*] from the horizon, and the other at $37\frac{1}{2}°$ [5 *punti*], strike the ground at equal distances; as do those shot at 60° and 30°, or at $67\frac{1}{2}°$ and $22\frac{1}{2}°$, and so on. Now let us hear the proof.

Commentary on *Two New Sciences*, pp. 245–246

It is interesting in the above paragraph to see Galileo's pride in having dealt with this issue in such a manner that experiment is unnecessary! Galileo offers a mathematical proof of his claim at the end of the final paragraph and brings his remarkable book to a conclusion by providing a number of additional theorems regarding projectile motion.

Conclusion

Judging from the Galilean materials studied in this chapter, what conclusion should be drawn about whether Galileo was strongly oriented toward an empirical approach? Because of Galileo's great importance in the seventeenth century, in the history of physics, and in the history of scientific method, this is a crucial question.

Based on these materials, the following analysis seems justified as a first approximation to characterizing Galileo's involvement with experimental methods in his *Two New Sciences*. First of all, consider the question: Is the traditional view correct that experimentation was crucial for Galileo, whereas it was sadly neglected by Aristotle, who relied far too much on pure thought? In pursuing this question, we first of all looked at Galileo's treatment of the issue of whether the weight of a body determines its rate of fall. What we found is that Galileo dealt with this question chiefly by means of a *thought experiment* rather than some sort of laboratory experiment. We then looked at his work on pendulums, especially his famous experiment with three pendulums acting simultaneously. Careful examination of this experiment shows that, for all its elegance and flair, the experiment is not performable. Moreover, it seems essentially certain that Galileo never performed it.

Regarding his work with inclined planes, in particular, a key point is his argument that the final velocity for all bodies rolling down inclined planes of the same height, but of different angles of inclination, are equal. In this instance, although Galileo did not demonstrate this directly by means of any experiment, he did use an experimental set up involving a pendulum to give it substantial justification. What helped most to make his case, however, was his argument that bodies in falling acquire enough impetus to regain the height lost. Another feature of this suggests the nature of Galileo's genius. He realized that *direct* empirical observation would not provide him with all the results he needed in regard to the motion of bodies rolling down inclined planes. He finessed this problem by showing how performable experiments with pendulums could help him make a convincing argument for his claim that the final velocity of balls rolling down inclined planes depends on the height of fall. Another case somewhat like this consists of the observations he reports about the degree to which stakes are driven into the ground when hit by falling balls.

Regarding Galileo's contribution to the law of inertia: it does not seem that experiment was crucial in his arguments for this law. Recall, for example, his quasi-thought experiments in the cabin of a ship. Recall also that after determining the angle of launch for a projectile that will maximize the distance it will traverse, Galileo congratulated himself on *not* having to carry out experiments on this matter.

Thus the role of experiment in Galileo's science is more complex than one might have thought. Of course, it would be wrong to deny that he experimented or that experiment was important to his work. It does seem that a purely empiricist or inductivist reading of Galileo is very problematic. But this does not necessarily entail concluding that empirical information was unimportant for Galileo. The chapter on Newton contains a presentation of what is called the "hypothetico-deductive method." It can be suggested that the method Galileo used does conform to the hypothetico-deductive method and that this method allows one to give proper credit to both the empirical and the deductive components in his methodology.

Chapter 3

From Galileo to Newton

Introduction

Isaac Newton once remarked: "If I have seen further it is by standing on the shoulders of Giants."[1] Galileo was certainly one of those giants. Other late sixteenth- and seventeenth-century figures were also important in preparing the way for the grand Newtonian system. Chapter Three provides information on some of these other giants.

William Gilbert (1544–1603)

In 1600, the Englishman William Gilbert published his *De magnete magneticisque corporibus et de magno magnete tellure physiologia nova* (*New Physics of the Magnet and of Magnetic Bodies, and of the Big Magnet, the Earth*), the most important study of magnetism published up to that time. Although based in substantial measure on earlier studies of magnetism, this book received considerable attention in the early seventeenth century, Galileo and Kepler being among those who avidly read it. The book is rich in

---

[1] From Isaac Newton's letter of 5 February 1675[6] to Robert Hooke as given in H. W. Turnbull (ed.), *The Correspondence of Isaac Newton,* vol. 1 (Cambridge University Press, 1959), p. 416. Spelling has been slightly modernized. There are two interesting facts about this famous quotation. One is that it does not appear in either *The Oxford Dictionary of Quotations* or Bartlett's *Familiar Quotations.* Actually, only two of Newton's statements (the same two) made it into these hefty catalogues. The other comment is that this metaphor was not of Newton's creation; in fact, it originated centuries before Newton with Bernard of Chartres, as Robert K. Merton has shown in a book length study of the history of this quotation: *On the Shoulders of Giants: A Shandean Postscript* (Free Press, 1965).

observations and experiments, but Gilbert gave little attention to quantifying his results. He did, however, provide useful insights and suggestions; for example, he distinguished between magnetic and electrical attraction (the latter is exemplified by the attraction exerted by a rubbed piece of amber). Moreover, he argued that the Earth is itself a large magnet. Gilbert also maintained that the Earth rotates but stopped short of advocating the Copernican doctrine of the Earth's annual revolution around the Sun. One should not attribute too much modernity to Gilbert. It is noteworthy, for example, that his theory of magnetism is ultimately animistic; he attributes a soul to the magnet, which when in the proximity of iron feels a desire for union with it.

Johannes Kepler (1571–1630)

Johannes Kepler was one of the most knowledgeable and creative astronomers of the early seventeenth century. He is now most widely remembered for what are called the Keplerian laws of planetary motion. Kepler presented the first two of these three laws in his *Astronomia nova* (*New Astronomy*). The third law appeared in Kepler's *Harmonices mundi* (*Harmony of the World*), which he published in 1619. It is important to keep in mind that these laws only gradually attained the prominence they now have in astronomy. Kepler did not himself number them or label them laws, nor did most of his contemporaries see them as much more than interesting conjectures. Moreover, Kepler worked out his first two laws primarily for the planet Mars, extending them, mostly by implication, to all the planets. A decade after presenting his first and second laws in 1609, Kepler attempted in his *Epitome astronomiae copernicanae* (*Epitome of Copernican Astronomy,* 3 vols., 1617–1621) to show their applicability to the other planets.

Kepler's First Law (which he actually arrived at second) is that planets move in ellipses with the Sun at one focus of the ellipse. An elementary knowledge of the geometry of ellipses contributes to an understanding of this conjecture. By Kepler's definition, an ellipse is the locus of points such that the sum of the distances of those points to two fixed points F_1 and F_2 (the foci of the ellipse) is constant. As is evident from the figure, for any point P on the ellipse, the sum of the lengths $F_1P + F_2P$ is constant. Consequently, an ellipse can be constructed by fixing the ends of a string to points F_1 and F_2 and moving a pencil at the end of the loop thus formed. The major axis of the ellipse is the line AB, the line extending through the foci to the opposite sides of the ellipse. The eccentricity (e) of an ellipse is the ratio of the distance F_1C to CA, which is also equal to F_1F_2/AB. The eccentricity ranges between

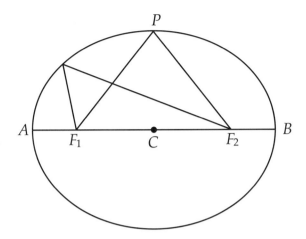

0, in which case the ellipse becomes a circle, and 1, in which case the figure reduces to a straight line. Kepler's First Law is, then, that any planet moves in an orbit of elliptical shape with the Sun at one focus and nothing at the other focus.

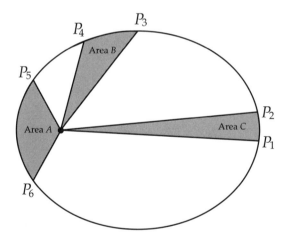

Kepler's Second Law is that a planet moves on its ellipse in such a way that a line from the Sun to the planet sweeps out equal areas in equal times. In the figure provided, if areas A, B, and C are equal, then the time taken by the planet in moving from P_1 to P_2 is equal to the time taken in moving from P_3 to P_4 or from P_5 to P_6. In this context it is interesting to ask where the planet moves most rapidly, where slowest.

In 1619, Kepler published his *Harmonices mundi*, in which he presented an array of harmonies or aesthetically pleasing relationships that he had detected in the universe. We need concern ourselves with only one of these, which is that if the time (T_P) taken by some planet P in moving through one complete orbital cycle (its sidereal period) is squared, and the distance (D_P) of the planet from the Sun is cubed, then for any planet P, Kepler's relationship can be expressed as

$$\frac{T_P{}^2}{D_P{}^3} = C,$$

where C is a constant that is identical for every planet. The power and importance of this law is suggested by the fact that if one knows the period and distance from the Sun of any planet (say, Earth, for which the period $T = 1$ year and $D = 1$ unit), then if one knows the sidereal period of any other planet, one can easily calculate the latter planet's relative distance from the Sun.

It is interesting to test the correctness of this relationship, which later became known as Kepler's Third Law, by checking it against the data given below.

Planet	Sidereal Period (in years)	Distance from Sun (normed on the earth as 1)
Mercury	.241	.387
Venus	.615	.723
Earth	1.000	1.000
Mars	1.881	1.524
Jupiter	11.862	5.203
Saturn	29.457	9.539

The importance of Kepler's *Astronomia nova* of 1609 derives not only from the two very precise mathematical laws he presented in it, but also because Kepler attempted a causal explanation of why the planets move as they do. Kepler's concern for causal considerations is evident from the full title of his book: *Astronomia nova* αἰτιολογητός, *seu physica coelestis,* which can be translated as *The New Astronomy Treated in Terms of Causes, or Celestial Physics.* Kepler not only sought a causal explanation of planetary motion, he no less significantly gave up the earlier idea of explaining these motions by means of the presence within the planet of an *anima* or some sort of intelligence that guides the planet through its path. Instead, Kepler formulated an explanation in terms of his idea of *vis,* which can be translated as force. Influenced by William Gilbert's notion that the Earth is a magnet, Kepler proposed that an immaterial *species* emanates from the

Sun in the plane of the ecliptic. Moreover, postulating that the Sun rotates on an axis perpendicular to the plane in which the planets move (the fact of the Sun's rotation was discovered only later), Kepler claimed that the rotation of the Sun gives a rotary motion to the *species,* which in turn causes the planets to rotate and revolve about the Sun. It is noteworthy that Kepler did not view this effect as decreasing inversely as the distance squared (as Newton asserted occurs for gravity), but rather that it decreases in inverse proportion to the distance. At double the distance, the effect is half as great. To put this in perspective, Kepler was searching for causal or in a sense mechanical (as opposed to animistic) explanations and in doing so, developed a notion that significantly contributed to the search for explanation in terms of gravitation.

René Descartes (1596–1650)

René Descartes and Isaac Newton created the two most influential seventeenth-century systems of physics. The Cartesian system (named after René Descartes or in its latinized form Renatus Cartesius) appeared chiefly in his *Principia philosophiae* [*Principles of Philosophy*], published in 1644, and in his *Le monde, ou traité de la lumière* [*The World, or Treatise on Light*], largely completed by 1633 but first published only in 1664. In these and his other physical works, Descartes set out a more comprehensive system of physical science than any that had appeared in the centuries since Aristotle. Descartes's system strongly influenced the system presented in 1687 by Isaac Newton in his *Philosophiae naturalis principia mathematica,* a work that contains many criticisms of the Cartesian system.

Descartes's brilliance in the mathematical sciences became apparent at least as early as 1637, when he published his *Discours de la méthode* and successfully sought to illustrate the power of his method by appending to that work three treatises: (1) *La géométrie,* which is the classic presentation of analytical or Cartesian geometry, (2) *La dioptrique,* the chief gem of which is Descartes's proof (the first ever published) of the law of refraction, and (3) *Les météores,* celebrated for contributing crucial insights into the optics of rainbows. There can be no question as to the importance of the new results that appeared in these three treatises, although scholars have questioned whether these results were in fact attained by the method Descartes had advocated.

Cartesian physics is substantially different not only in content but also in method from that later advocated by Newton. One example of this will be evident in the selections from Descartes that follow. This is that although

Descartes repeatedly stressed the power and importance of mathematics, his physical writings contain rather less use of mathematics than such statements would lead one to expect. Moreover, Descartes employed in his physical writings much less mathematics than either Galileo had done or than Newton would later do. Indeed, Newton in choosing the title for his *Principia* very possibly added the modifier *mathematica* to set off his heavily mathematical and frequently quantified discussions from the more qualitative presentations of his French predecessor.

One reason that Descartes employed less mathematics than Galileo was that Descartes's approach was significantly different. This is evident from a criticism Descartes leveled at Galileo's *Two New Sciences* in a letter to Mersenne of 11 October 1638, in which Descartes complained that Galileo has not "considered the first causes of nature [but] has only sought the reasons of some particular effects, and thus he has built without foundation." And he added: "Everything he says about the speeds of bodies descending in the void, etc., is built without foundation, for first he should have determined what gravity [*pesanteur*] is. . . ."[2] What Descartes meant by this was that Galileo had concentrated on *how* bodies move, leaving aside the question of *why* they move, which Descartes took to be at least as important. Moreover, Descartes dealt in general with a broader range of questions than Galileo—for example, the structure of the universe as a whole. Thus it is not surprising that Descartes in large part failed to recognize the importance of the work of Galileo. The great Italian physicist frequently looked to empirical phenomena in hopes of detecting the laws of nature, whereas Descartes looked within his mind to find possible laws of nature and then checked these possible laws against the actual phenomena in order to see which of the possible laws God in fact chose to put into operation.

The differences should not, however, be overstressed; Galileo to some extent, Descartes and Newton to a far greater degree, shared a commitment to what has been described as the **mechanical philosophy**, the view that aspects of physical nature must be explained not in terms of spirits, souls, natures, or occult qualities, but rather in mechanical terms. Christiaan Huygens provided a classic statement of this point of view when he wrote in his *Treatise on Light* (1690) that "the true Philosophy [is that] in which one conceives the causes of all natural effects in terms of mechanical motions.

---

[2] As translated in P. Damerow, G. Freudenthal, P. McLaughlin, and J. Renn, *Exploring the Limits of Preclassical Mechanics: A Study of Conceptual Development in Early Modern Science: Free Fall and Compounded Motion in the Work of Descartes, Galileo, and Beeckman* (Springer-Verlag, 1991), p. 333. For the French text, see *Oeuvres de Descartes,* ed. by Charles Adam and Paul Tannery, vol. 2: *Correspondance* (Libraire philosophique J. Vrin, 1969), pp. 380, 385.

This, in my opinion, we must necessarily do, or else renounce all hopes of ever comprehending anything in Physics."[3] Galileo's famous distinction in his *Assayer* (1623) between **primary qualities** (size, shape, location in space and time, or motion, which qualities are all quantifiable) and **secondary qualities** (color, temperature, taste, or sound, which qualities do not, strictly speaking, exist in nature), was a central idea in mechanical philosophy, and one that suggests that the roots of the mechanical philosophy extend back to the atomistic philosophies of antiquity. Descartes's strong distinction between mental and physical substances was also part of this program, as was his attempt to explain many phenomena of the universe in terms of a subtle all-pervasive matter.

Descartes's discussion of the solar system in his *Principia philosophiae* provides a good idea of his approach. Descartes states:

> . . . let us assume that the matter of the heaven, in which the Planets are situated, unceasingly revolves, like a vortex having the Sun at its center, and those of its parts which are close to the Sun move more quickly than those further away; and that all the Planets (among which we [shall from now on] include the Earth) always remain suspended among the same parts of this heavenly matter. For by that alone, and without any other devices, all their phenomena are very easily understood. Thus, if some straws [or other light bodies] are floating in the eddy of a river, where the water doubles back on itself and forms a vortex as it swirls: we can see that it carries them along and makes them move in circles with it. Further, we can often see that some of these straws rotate about their own centers, and that those which are closer to the center of the vortex which contains them complete their circle more rapidly than those which are further away from it. Finally, we see that, although these whirlpools always attempt a circular motion, they practically never describe perfect circles, but sometime become too great in width or in length, [so that all the parts of the circumference which they describe are not equidistant from the center]. Thus we can easily imagine all the same things happen to the Planets; and this is all we need to explain all their remaining phenomena.[4]

One can get a fuller idea of what Descartes had in mind by examining a diagram he provided in his *Le Monde* (see page 93 below). There the letter

---

[3]Christiaan Huygens, *Treatise on Light,* trans. by Silvanus P. Thompson (Dover Publications, n.d.), p. 3.

[4]As quoted in Julian B. Barbour, *Absolute or Relative Motion? A Study from a Machian Point of View of the Discovery and the Structure of Dynamical Theories,* vol. I: *The Discovery of Dynamics* (Cambridge University Press, 1989), p. 415. For the Latin text, see *Oeuvres de Descartes,* ed. by Charles Adam and Paul Tannery, vol. 8:1: *Principia philosophiae* (Libraire philosophique J. Vrin, 1964), p. 92.

S represents the Sun. It is surrounded by the planets, which move in the vortex centered on the Sun. The Earth is designated by T (Terra).

Other vortices are centered on other stars (at points E and A, for example). Near the top of the diagram is a double line representing the path of a comet, which body Descartes conceived to be larger than the planets and hence capable of moving from one vortex to another. The matter at the center of the vortex, i.e., in the Sun, consists, according to Descartes, of finely divided particles moving at very high speeds, whereas as one proceeds farther out from the Sun, the particles filling space are larger, heavier, and move more slowly. Descartes interpreted light as a pressure coming from the Sun and extending out to the planets. Because it consists of pressure, it is propagated instantaneously. A very important part of Descartes's conception is his claim that there can be no empty space, that a vacuum free of all matter is an impossibility. Another characteristic feature of Descartes's view is that what is known as "action at a distance" is impossible. What this means is that no body can act on another body unless it is somehow in direct physical contact with it. Within this perspective, the idea of, say, the Earth exerting a gravitational attraction on the Moon through the empty intervening space is not acceptable.

Perhaps the most important contribution made to physics by René Descartes was his formulation of the law of inertia, which appeared in 1644 in his *Principia philosophiae.* In the same work, Descartes also proposed the important idea that the quantity of motion in the universe is constant. Descartes went so far as to claim that when God created the universe, God gave it a certain quantity of motion, which quantity has remained unaltered since that time. Both these ideas are developed in the selection that follows from Descartes's *Principia philosophiae.* In reading this selection, it is useful to ask: What role did the idea of the relativity of motion play in Descartes's formulation of the law of inertia?

Descartes, *Principles of Philosophy*, pp. 54–66[5]

26. No more action is required for motion than for rest.

Indeed, it must be noted that we are laboring under a great prejudice, in that we consider that more action is required for motion than

---

[5]Selection from René Descartes, *Principles of Philosophy,* translated by William H. Donahue from the Latin text given in Charles Adam and Paul Tannery (eds.), *Oeuvres de Descartes,* vol. 8.1 (Libraire Philosophique J. Vrin, 1973), pp. 54–66. There is also a French translation (in vol. 9 of the same edition) that was made with Descartes's approval. Material from the French has occasionally been introduced into the present version. These instances are always noted in the footnotes.

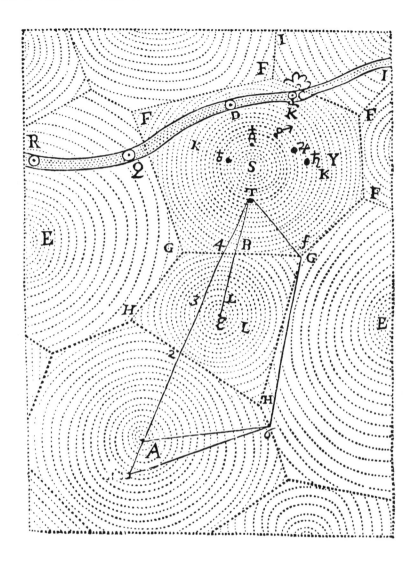

Descartes's vortices, from *Le Monde* (1677)

for rest. And we have persuaded ourselves of this from our youngest age, because our body is usually moved by our will, of which we are inwardly conscious, and it rests for the sole reason that it adheres to the Earth through its heaviness, the force of which we do not perceive. And because that heaviness, and many other causes not noticed by us, resist the motions that we produce in our members by our will, and make us tired, we think that there is a need for more action or more force in producing motion than in stopping it. That is, we take the action as being the effort we exert in moving our members, and moving other bodies by means of them. We will nonetheless easily cast off this prejudice if we consider that we need effort not only for moving external bodies, but frequently for stopping their motions as well, when they are not stopped by heaviness or some other cause. As, for example, we do not make use of a greater action (or at least not much more) in setting a ship in motion that is at rest in quiet water, than in abruptly holding back the same ship when it is in motion. And if experience makes us see in this instance that a little less is needed to stop it than to make it go, it is because the heaviness and the viscosity of the water displaced by it when it is in motion, by which it might be gradually brought to a stop, must be removed.[6]

27. Motion and rest are only different modes of the moved body.

Moreover, what is in question here is not the action that is understood as being in the mover or in that which brings the motion to a stop, but only transport, or the absence of transport (or rest). Therefore, it is obvious that this transport cannot exist outside the body that is moved, and that this body is in one mode when it is being transported and in another when it is not being transported or when it is at rest, so that motion and rest are nothing but two different modes in the body.

28. Motion, taken in its proper sense, is in relation only to the bodies contiguous to the moved body.

I have added, in addition, that transport takes place *from the neighborhood of contiguous bodies to the neighborhood of others,* and not from one place to another. For, as I have explained above, "place" is taken in a variety of ways, and depends on our thinking. But since by "motion" we understand that transport that takes place away from the neighborhood of contiguous bodies, and since only one set of bodies can be contiguous to the same body in the same moment of time, we cannot attribute many motions to that body in the same time,.but one only.

29. Motion is also in relation only to those contiguous bodies that are seen as being at rest.

And finally, I have added that the transport takes place, not from the neighborhood of just any contiguous bodies, but only *of those that are seen as being at rest.* For transport is itself reciprocal, and body *A B* cannot be understood to be moved from the neighborhood of body

---

[6]For the sake of increased clarity, some material from the French version (see footnote 5 above) has been added in the last sentence. [translator's note]

CD without the body CD being at the same time understood to be moved from the neighborhood of body AB; and furthermore, exactly the same power and action is required in the one case as in the other. For this reason, if we were to attribute to motion a nature that belongs to it alone, and is not related to another motion, we would be saying that when two contiguous bodies are in motion, one in one direction and the other in the other direction, and are thus separated from each other, there would be as much motion in the one as in the other. But this is excessively at variance with common usage in speaking; for since we are accustomed to standing on the Earth, and to considering it as at rest, even though we see certain parts of the Earth, contiguous to other smaller bodies, moved away from their neighborhood, we do not therefore consider the Earth itself to move.

...

36. God is the primary cause of motion, and always conserves the same quantity of motion in the universe.

Now that the nature of motion has been discussed in this way, it is appropriate to consider its cause. This is twofold; namely, first, the universal and primary cause, which is the general cause of all motions that exist in the world; and second, the particular cause, by which it happens that individual parts of matter acquire motions that previously they had not had. And as for the general cause, it appears obvious to me that this is nothing other than God Himself, who at the beginning created matter at the same time with motion and rest, and who now, solely through His ordinary concurrence, preserves that same amount of motion and rest in that matter as a whole as He placed in it then. For even though in the moved matter that motion is nothing other than its mode, it nevertheless has a certain and determinate quantity, which we easily understand to be able to be the same in the whole universe of things, although it may change in its individual parts. Thus, for example, we consider that when one part of matter moves twice as fast as another, and this latter is twice as great as the former, there is the same amount of motion in the smaller as in the greater; and the slower the motion gets in one part, the faster the motion gets in some other part equal to it. We also understand that in God there is the perfection, not only that He is unchangeable in Himself, but also that He works in a manner that is as far as possible constant and unchanging; so much so that, except for those alterations that evident experience or divine revelation make certain, and that we perceive or believe to occur without any change in the Creator, we ought not to suppose any other alterations in His actions, lest this be used as evidence of inconstancy in Him. As a consequence, it is most highly in agreement with reason for us to think that from the sole fact that God had moved the parts of matter in various ways when He first created them, and that He now should preserve all that matter in exactly the same mode and in the same proportion in which He first created it, He also ceaselessly conserves the same amount of motion in that matter.[7]

37. The first law of nature: that each thing (insofar as this depends on itself), persists in the same state, and thus what is once moved always carries on moving.

And from this same unchangeability of God, there can be known certain rules or laws of nature, which are secondary and particular causes of the diverse motions that we notice in individual bodies. The first of these is that any one body, insofar as it is simple and undivided, always remains in the same state, insofar as this depends on itself, and does not ever change except through external causes. Thus, if some part of matter be a square, we easily persuade ourselves that it is forever going to stay square unless something should come from elsewhere that would change its shape. If it is at rest, we do not believe that it is ever going to get moving unless it be impelled to do so by some cause. And there is no greater reason why, if it be in motion, we should think that of its own accord, and not hindered by anything else, that motion that it has is ever going to be discontinued. And therefore it must be concluded that whatever moves always

---

[7]This is one of the most important parts of Descartes's book. Ever since the publication of Descartes's ideas that God created a certain definite quantity of motion in the universe and that, moreover, God set down as a law of nature that this quantity of motion must remain unchanged, scientists, theologians, and philosophers have debated what this might mean and whether it can be true. Not surprisingly, much of the debate centered on how the quantity of motion should be measured. Although Descartes never stated precisely how his quantity of motion should be measured, the usual interpretation is that by a body's quantity of motion, he meant to include both its (scalar) speed and the quantity of matter it contains. Some other candidates for measuring the quantity of motion that were put forward include simply its speed, its (vectorial) velocity, or its velocity times its weight or mass (what we now call "momentum"). A particularly noteworthy candidate that began to emerge during the latter half of the seventeenth century and that attracted increasing attention in the nineteenth century was $\frac{mv^2}{2}$, which came to be called kinetic energy. It is also worth noting at this point that the question of the conservation of the quantity of motion has important implications in physics. For example, scientists attempted to analyze what happens when bodies collide in terms of the conservation of the bodies' quantities of motion. [MJC]

moves, as much as it can. But because we are situated here around the Earth, whose physical arrangement is such that all motions that take place near it soon cease, and often through causes that escape our senses, therefore, from our youngest age, we have often judged that those motions that have ceased through causes unknown to us, stopped on their own. And now we are inclined to consider what we see to have been experienced in many cases to be true of everything, namely, that they stop by their own nature, or have a tendency to rest. This is certainly opposed to the laws of nature in the highest degree, for rest is the contrary of motion, and nothing can through its own nature be carried over to its contrary, that is, to the destruction of its own self.

38. Why bodies pushed by the hand continue to move after it has left them.[8]

And, in fact, everyday experience of things that are thrown fully confirms our rule. For there is no reason why projectiles should continue in motion for some time after they have been separated from the throwing hand other than that once they have been moved they continue to move until they are slowed by bodies in their way. And it is obvious that they are usually slowed gradually by air or by whatever other fluid bodies in which they move, and that therefore their motion cannot last a long time. For by the sense of touch we can experience that air resists the motion of other bodies, if we strike it with a fan; and the flight of birds confirms the same thing. And there is no other liquid that would not resist the motion of projectiles more obviously than air.

39. The second law of nature: that all motion is in itself straight, and therefore the things that move circularly always tend to recede from the center of the circle that they describe.

The second law of nature is: that any part of matter whatever, considered in itself, does not ever tend to continue its motion along any oblique lines, but only along straight ones. However, many parts are often compelled to turn aside because others get in the way, and (as was said a little earlier)[9] in any given motion a circle is made (to speak loosely), from all matter moving at the same time. The

---

[8]This heading is from the French version (see footnote 5 above). The original Latin reads simply, "On the motion of projectiles." [translator's note]
[9]Article 33. [translator's note]

cause of this rule is the same as that of the preceding one, namely, the unchangeability and simplicity of working, through which God conserves motion in matter. For He conserves it precisely as it is only in that very moment of time in which He conserves it, without taking any account of the state that happened to exist a little earlier. And although no motion happens in an instant, it is nonetheless manifest that every thing that moves, in the individual instants that can be noted while it moves, is determined to continue its motion in some direction along a straight line, and not ever along any curved line. As, for example, a stone A in a sling EA rotated on circle ABF, at the instant when it is at point A, is truly determined to

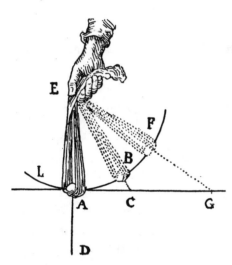

[carry out] a motion in a certain direction, namely, along a straight line toward C, such that the straight line AC is tangent to the circle. Moreover, one cannot pretend that it is determined to [carry out] any curved motion, for even though it might previously have come from L to A along a curved line, nevertheless, nothing of that curvature can be understood to remain in it when it is at the point A. And this is confirmed by experience, because if it should then come

forth out of the sling, it will not carry on to move toward B, but toward C. From this it follows that every body that moves circularly always tends to recede from the center of the circle that it describes. This we experience by the very sense of our hand in [the case of] a stone, when we swing it around with a sling.[10] And because we will make frequent use of this consideration in what follows, it is to be noted carefully, and will be set forth at greater length below.[11]

40. The third law: that one body, encountering another stronger one, would not give up any of its motion; whereas in encountering one that is less strong, it would give up as much as it transfers to that one.

---

[10]The French version continues, "for the stone pulls on the string and stretches it in an effort to remove itself directly away from our hand." [translator's note]

[11]Part III, articles 57 and 58. [translator's note]

The third law of nature is this: where a body that is in motion encounters another, if it should have less power [*vis*] to carry on along a straight line than this other one has to resist it, it is then turned away in the other direction, and, keeping its motion, gives up only the directedness [*determinatio*] of the motion; if, on the other hand, it should have a greater power, then it moves the other body along with itself, and it loses as much motion as it gives to the other from its motion. Thus we know by experience that when any hard bodies whatever are propelled, and strike upon another hard body, they do not as a result stop moving, but are reflected back in the opposite direction. When, on the other hand, they encounter a soft body, they are immediately brought to a stop because they easily transmit all their motion to it. And all the particular causes of the changes that happen to bodies are contained in this third law—at least those [causes] that are themselves corporeal, for whether, and how, human or Angelic minds may have the power to move bodies, we are not now investigating, but we reserve this for the *Treatise on Man*.

41. Proof of the first part of this law.

Now the first part of this law is demonstrated from the difference that arises between motion seen in itself and its being destined towards a certain direction, whereby that destination can be changed while the motion remains entire. For, as was said previously [law 1], any thing whatever that is not a compound, but is simple, as motion is, will always continue to exist, as long as it is not destroyed by some external cause. And in an encounter with a hard body, there does indeed appear a cause that would prevent the motion of the other body that it encounters being directed towards the same side, but no cause at all that would remove or diminish the motion itself, because motion is not contrary to motion. Therefore it follows that for these reasons it ought not to be diminished.[12]

42. Proof of the second part.

Further, the second part is proved from the immutability of God's working, continually preserving the world now by means of the same action by which He once created it. For because everything is filled

---

[12] A simple example of what Descartes has in mind in this discussion would be the bounce of a tennis ball against the Earth's surface. The ball is so small, the Earth so large, that the quantity of motion of the Earth will not be appreciably changed by this impact. What happens with the ball is that its quantity of motion is not changed, but the direction (Descartes uses the word "destination") of that motion is altered. If the ball is dropped perpendicularly to the Earth, the direction of its motion is exactly reversed. [MJC]

with bodies, and the motion of any particular body nonetheless tends into a straight line, it is clear that from the beginning, in creating the world, God not only moved different parts of it in different ways, but also, and at the same time, brought it about that certain parts impelled other parts, and transferred their motions to them. In exactly the same way, preserving the world now by the same action and by the same laws with which He created it, He preserves motion, not always instilled in the same parts of matter, but passing over from some of them into others in accordance with their mutual encounters. And thus this continual change of created things is itself evidence of the immutability of God.

Christiaan Huygens (1629–1695)

During the latter half of the seventeenth century, the leading Continental contributor to mechanics was the Dutch scientist Christiaan Huygens, who also made major contributions to astronomy and to optics. Huygens was born at the Hague in Holland, the son of Constantin Huygens (1596–1687), a prominent diplomat, poet, and composer, whose home attracted such visitors as René Descartes. In fact, Descartes's ideas strongly influenced the young Huygens. Another significant influence was Pierre Gassendi, whose orientation was Epicurean or atomistic. In 1645, Huygens entered Leiden University, where he studied mathematics and law. Then, in 1647, he began a two year period of studying law at the Collegium Arausiacum (College of Orange) at Breda. Among his earliest major scientific achievements were his discovery of the largest Saturnian moon, Titan, in 1655, and the invention of the pendulum clock in 1657. Huygens's pendulum clock was based on Galileo's discovery of the isochronism of the pendulum. The mechanics of pendulum clocks was the subject of one of his most important books, his *Horologium oscillatorium, sive de motu pendulatorum* (*The Pendulum Clock, or the Motion of Pendulums*), which he published in 1673. In that volume, he presented the law of centripetal[13] acceleration, which is one of the fundamental laws of mechanics. It will be discussed shortly. In 1663, while in England, Huygens attended some meetings of that country's newly founded (1660; royal charter 1662) Royal Society, which elected him a Foreign Member. Three years later, Huygens became a member of the Académie des sciences in Paris. He spent most of the next fifteen years as a salaried member of the Académie before returning to Holland on a permanent basis

---

[13]Centripetal in relation to circular motion means toward the center; centrifugal means away from the center.

in 1681. In 1672 in Paris, Huygens met the young Gottfried Wilhelm Leibniz (1646–1716), who studied mathematics with him for a period. Two years later Huygens transmitted Leibniz's first paper on the differential calculus to the Académie des sciences.

It was in 1679 that Huygens presented the first draft of what ranks as his most important contribution to physics; this was his *Traité de la lumière* (*Treatise on Light*), which was finally published in 1690 and which is usually seen as the source of the wave theory of light. Huygens's main rival during this period in regard to the theory of light was Isaac Newton, whose great work in the theory of light, his *Opticks,* was published in 1704, although many of the key ideas in that book had been made public earlier. Newton and Huygens met in person only once, which occurred in 1689 when Huygens traveled to England.

Huygens and the Law of Centripetal Acceleration

Possibly Huygens's single most important contribution to mechanics was what is now called the **law of centripetal acceleration**.   Actually the term centripetal was coined by Newton, who introduced it in contrast to the term centrifugal, which was Huygens's creation. Huygens believed he was finding a measure of the *conatus* or "striving" of a body to recede from the center when it was deflected from its straight path and forced to move in a circle. The law is

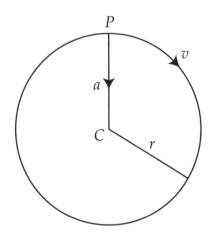

that for a particle P moving on a circle of radius r at a constant speed v, the acceleration of the particle towards the center of the circle equals its speed squared divided by the radius of the circle. The formula that results is: $a = \frac{v^2}{r}$. To understand its meaning, consider the particle moving in its circular path. Using the distinction between speed, which is how rapidly a body is moving, and velocity, which refers to its speed in a particular direction, it can be suggested that although the speed of the body is constant, its velocity is changing. What is the direction of that change in motion? It can be shown that a very effective way to analyze this situation is to attribute to the particle a uniform acceleration toward the center of the circle. Newton will later take this approach in analyzing the motion of the Moon. Newton claims that the path of the Moon is the result of (1) the velocity the Moon has at any given time, which at any instant is in a direction tangent to the curved path of the Moon, and (2) the tendency of the Moon to move (in effect, to fall) toward the Earth because of gravitational attraction. The law, however, is more general than this. It makes no reference to forces, for example. It is noteworthy that, in his 1673 publication of this law,[14] Huygens did not include a derivation of the law. His derivation became known only when it was published in 1703 in Huygens's *Opera Posthuma*. It is also significant to note that Huygens did not suggest applying this law to the planets, as Newton was to do with great success.

Huygens and the Theory of Collisions

What paths do bodies, for example, billiard balls, take after colliding with each other? This question was widely discussed in the mid-seventeenth century, Descartes having produced one of the most famous theories. Some parts of his theory appear in the selection from his *Principles of Philosophy* given in this chapter. By at least the late 1660s, Descartes's laws of impact had come to be questioned. In 1668, England's Royal Society issued an invitation for scientists to formulate more satisfactory laws. In response to that challenge, Huygens, who had already worked out new laws for collisions between such "hard bodies" as billiard balls, submitted his results. The same laws had also been worked out by Christopher Wren (1632–1723), although Huygens provided a derivation of the laws from general principles. In 1669, Huygens's results were published in both the Royal Society's *Transactions* and in the *Journal des sçavans,* but only in an abbreviated form. His more thorough analysis appeared only in 1703 under the title *De motu corporum ex percussione* (*On the Motion of Bodies in Collisions*). The opening portion of that work is given next. One reason for including this selection is that it shows how brilliantly Huygens had thought about and made use of the idea of the relativity of motion.

In the short selection provided, Huygens works out what must happen when, say, a stationary billiard ball is hit squarely by an identical ball moving toward it. In resolving this question, Huygens not only attained a correct result, but also proved his results by an elegant and effective method.

---

[14]For a translation of the relevant section of Huygens's *Horologium oscillatorium,* see Christiaan Huygens, *The Pendulum Clock or Geometrical Demonstrations Concerning the Motion of Pendula as Applied to Clocks,* trans. by Richard Blackwell (Iowa State University Press, 1986), pp. 176–178.

Selection from Huygens, *On Colliding Bodies*[15]

Hypothesis I: Any body already in motion will continue to move perpetually with the same speed and in a straight line unless it is impeded.

Hypothesis II: Whatever may be the cause of hard bodies rebounding from mutual contact when they collide with one another, let us suppose that when two bodies, equal to each other and having equal speed, directly collide with one another, each rebounds with the same speed which it had before the collision.

Two bodies are said to directly collide when both their motion and their contact occur in the same straight line which passes through the center of gravity of each body.

Hypothesis III: Both the motion of bodies and their equal or unequal speeds must be understood in relation to other bodies considered to be at rest, even if both sets of bodies happen to be involved in some other common motion. As a result, when two bodies collide, then even if each of them is simultaneously subject to some other additional equal motion, they will in no way act on each other with respect to the common motion by which they are each moved. It is as if that additional motion were totally absent.

For example, someone who is carried along by a boat which progresses with uniform speed makes two equal balls collide with each other with equal speed as determined in relation to himself and to the parts of the boat. We say then that each ball ought to rebound with an equal speed in relation to the man carried along in the boat, as would

[15]From "Christiaan Huygens' *The Motion of Colliding Bodies*," trans. by Richard J. Blackwell, *Isis, 68* (1977), 574–597:574–576.

clearly also happen if he had made the same balls collide with equal speed while he was standing in a motionless boat or on the ground.

Having made these suppositions concerning the collision of equal bodies, we will demonstrate the laws by which they mutually collide; we will then proceed to infer in the proper place other hypotheses which we will need for the case of unequal bodies.

Proposition 1: If a body collides with another equal body which is at rest, then after the contact the former is at rest and the latter acquires the same speed which was in the body which struck it.

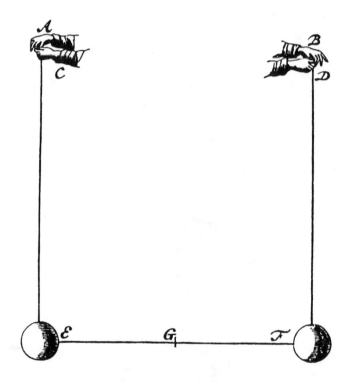

Imagine a boat which is floating down river parallel to the bank and so close to the bank that a sailor standing in it can extend his hands to a friend standing on the bank. The sailor holds in his hands, *A* and *B*, two bodies, *E* and *F*, suspended on strings. The distance *EF* is divided in half at the point *G*. As he moves his hands together into contact with equal motions as judged in relation to himself and to the boat, at the same time he makes the balls *E* and *F* collide with equal speed. Thus it is necessary that the balls rebound from mutual contact with equal speed [Hypothesis II] in relation to the sailor and the boat. Moreover it is supposed that the boat meanwhile has moved toward the left with the speed GE which is the same speed by which the left hand A was moved toward the right.

From this it is clear that the sailor's hand A was unmoved in relation to the bank and to his friend standing there, and also that the sailor's hand B was moved in respect to his friend with the speed FE which is double that of GE or FG. Accordingly let us suppose that the friend standing on the bank takes hold of, with his own hand C, the sailor's hand A together with the top of the string from which the ball E is suspended; also he takes hold of, with his hand D, the sailor's hand B which holds the string from which F is suspended. It is apparent then that while the sailor made the balls E and F collide with equal speed as judged in relation to himself and the boat, at the same time his friend standing on the bank made the moving ball F collide with the motionless ball E with a speed FE as judged in relation to himself and to the bank. It is also clear that the sailor who moves the balls as described is not affected in any way by the fact that his friend standing on the bank has taken hold of his hands and the tops of the strings since the motion of the friend's hands is only a companion motion and produces no hindrance for the sailor. For the same reason the friend on the bank who moves the ball F toward the motionless ball E is not affected by the fact that his hands and the sailor's hands are joined, since both hands A and C are at rest in relation to the friend and the bank and both hands B and D are moved with the same speed FE. However, as was said, after mutual contact the balls E and F rebound with equal speed in relation to the sailor and the boat, the ball F with speed GE and the ball E with speed GF, and all the while the boat itself moves toward the left with the speed GE or FG. Hence it follows that in relation to the bank and the friend standing there, the ball F comes to rest after the collision and the other ball E moves toward the left with a speed of double GE, that is, the speed FE which is the same speed by which the ball F was originally moved toward the ball E. Thus we have shown that when a man standing on the ground makes a body collide with another equal motionless body, then after the contact the former body loses all its motion and the latter acquires all of it.   Q.E.D.

Commentary

It is easy to assign names to Huygens's three "hypotheses." The first is simply a version of the law of inertia, or more precisely, a special case under it. The second can be labeled Huygens's symmetry hypothesis. The third has been designated Huygens's Principle of the Relativity of Motion.[16] Various authors have seen this principle as an important step towards Einstein's theory of relativity.[17] Although the principle was widely recognized in the seventeenth century, it is significant that Huygens demonstrated, probably more fully than anyone else in this period, including Descartes, the power the principle packed for investigations in physics.

Another feature to be noted about Huygens's demonstration is that already at this point he had shown that Descartes's rules for collisions contain flaws. Descartes in his *Principles of Philosophy* had discussed the question treated by Huygens in Proposition 1: What is the motion produced when a stationary ball is struck squarely by an identical ball hitting it with some particular velocity? Descartes concluded (his sixth rule) that both balls will be driven off in opposite directions, the moving ball with three times the speed of the formerly stationary ball.[18] Significant as Huygens's correction is, it should not obscure a broader point, which is evident from comparing the texts of Descartes and Huygens. This point is that Huygens was almost certainly led to all three hypotheses mentioned in the above paragraph by reading Descartes, even if the Dutch scientist understood these ideas more fully and employed them more effectively than his French predecessor.

Huygens's *De motu corporum* of course carries the discussion much farther than this first proposition. His second proposition, which he derived from his three "hypotheses" in a very similar way by analyzing motions first as experienced in a uniformly moving boat, then from the point of view of the shore, states: "If two equal bodies moving with unequal speeds collide with each other, then after the contact they move with their speeds reciprocally exchanged." With his third proposition, Huygens opens a discussion of what happens when a smaller ball collides with a larger ball. This discussion continues through his formulation of eleven more propositions dealing with collisions among hard bodies.

One final important point about *De motu corporum* needs to be made. Huygens's mode of analysis of collisions differs in a major way from the mode of analysis that has come to dominate in mechanics. This modern method is to begin with the law of conservation of momentum (mv) and with the law of conservation of kinetic energy, which is that the quantity remains

---

[16] It is useful to compare Huygens's statement with Newton's Corollary 5 to his three laws of motion.

[17] Julian Barbour states: "*De Motu Corporum* is a pre-run of [Einstein's] 1905 paper that created special relativity. In fact, the editors of Huygens's works comment that if the principle of relativity should carry anyone's name, then it should be Huygens." See Barbour's *Absolute or Relative Motion?* vol. I, p. 457.

[18] René Descartes, *Principles of Philosophy*, trans. by Valentine R. Miller and Reese P. Miller (Reidel, 1983), 67–68.

constant. Huygens was not unaware of the significance of momentum and of what was in effect the quantity mv^2. In fact, in *De motu corporum* he discussed the validity and applicability of the traditional form of the conservation of motion (Proposition 6) and derived a form of the law of conservation of kinetic energy (Proposition 11).[19] Moreover, in his brief paper (mentioned above) published in 1669 in the *Journal des sçavans,* he provided statements of both the law of conservation of momentum and of kinetic energy. His statement of the momentum conservation law is "The quantity of motion that two bodies possess may either increase or decrease as a result of their collision; but the quantity towards the same side, the quantity of motion in the contrary direction having been subtracted, always remains the same."[20] In other words, he specified that Descartes's law of conservation of quantity of motion can be maintained provided that (in modern terms) the momentum is treated as a vectorial quantity. His statement of the law of conservation of kinetic energy is "The sum of the products of the size of each hard body multiplied by the square of its velocity is always the same before and after their collision."[21]

Huygens: Conclusion

Huygens deserves credit for presenting more clearly than anyone earlier a statement of what is now known as the hypothetico-deductive method. In his *Treatise on Light,* he stated:

> There will be seen in [this book] demonstrations of those kinds which do not produce as great a certitude as those of Geometry, and which even differ much therefrom, since whereas the Geometers prove their Propositions by fixed and incontestable Principles, here the Principles are verified by the conclusions to be drawn from them; the nature of these things not allowing of this being done otherwise. It is always possible to attain thereby to a degree of probability which very often is scarcely less than complete proof. To wit, when things which have been demonstrated by the Principles that have been assumed correspond perfectly to the phenomena which experiment has brought under observation; especially when there are a great number of them, and further, principally, when one can imagine and foresee new phenomena which ought to follow from the hypotheses which one employs, and when one finds that therein the fact corresponds to our

[19] H. J. M. Bos, "Christiaan Huygens," *Dictionary of Scientific Biography,* ed. by C. C. Gillispie, vol. 6 (Charles Scribner's Sons, 1972), pp. 597–613: 603.

[20] As given in Barbour, *Absolute or Relative Motion?* vol. 1, p. 472. See Christiaan Huygens, *Oeuvres complétes de Christiaan Huygens, 16* (Martinus Nijhoff, 1929), pp. 179–181:180.

[21] As given in Barbour, *Absolute or Relative Motion?* vol. 1, p. 472. See Huygens, *Oeuvres complétes, 16,* pp. 179–181:180.

prevision. But if all these proofs of probability are met with in that which I propose to discuss, as it seems to me they are, this ought to be very strong confirmation of the success of my inquiry; and it must be ill if the facts are not pretty much as I represent them.[22]

We shall return to this statement later for a fuller analysis. But now: Newton.

---

[22] Huygens, *Treatise on Light* (footnote 3), pp. vi–vii.

Isaac Newton

Chapter 4

Newton and Mechanics

Chronology of the Life of Sir Isaac Newton

1642   April: Marriage of the illiterate Isaac Newton, Sr., and the semi-literate Hannah Ayscough. October: death of Isaac Newton's father.

December 25:[1] birth of Isaac Newton at Woolsthorpe Manor in Lincolnshire in England.

1646   January 27: Hannah Newton, widow, marries 63 year old Barnabas Smith, a clergyman living nearby. Mother departs from Woolsthorpe, leaving her son Isaac to be raised by his grandmother.

1653   Death of Barnabas Smith, Newton's stepfather; Newton's mother returns to Woolsthorpe.

1658   September 3: Death of Oliver Cromwell.

1660   May 25: King Charles II returns to England.

1661   After studies at King's School in Grantham, where he receives a good start in mathematics from the schoolmaster John Stokes, Newton enters Trinity College of the University of Cambridge as a sub-sizar (a student required to work to defray his college costs).

1663–4   In late 1663 or early 1664, Newton begins reading Walter Charleton's *Physiologia Epicuro-Gassendo-Charltoniana* (1654), which leads Newton to adopt an atomistic view of the universe in opposition to the Aristotelian, Scholastic, and Cartesian cosmologies

---

[1] The calendars of persons on the continent, at least those in countries where the Gregorian Calendar of 1582 had already been adopted, read January 4 rather than December 25 at the time of Newton's birth, but the English, ever on guard against popery, had as yet resisted the Gregorian Reform of the calendar.

then available. During the latter half of 1664, Newton carefully reads Descartes's *Principia philosophiae* (*Principles of Philosophy*) of 1644, in which he finds much of interest.

1665   January: receives his B.A. degree.

1665–6   Because of the plague, Cambridge University closes from August, 1665 to March, 1666 and from June, 1666 to April, 1667, Newton returning to Woolsthorpe. The period from early 1665 to summer, 1666 is known as Newton's **annus mirabilis** (miraculous year) because he makes four major discoveries during that time.

1. **The Binomial Theorem:** "In the beginning of the year 1665 I found the Method of approximating series & the Rule for reducing any dignity [power] of any Binomial into such a series."[2]

2. **The Calculus:** "The same year [1665] in May I found the method of Tangents... & in November had the direct method of fluxions... and in May following I had entrance into y^e [the] inverse method of fluxions."[3]

3. **The heterogeneity of white light:** "and the next year [1666] in January had the Theory of Colours...."[4]

4. A first step toward the **theory of universal gravitation:** "And the same year [1666] I began to think of gravity extending to ye orb of the Moon & having found out how to estimate the force with w^{ch} [a] globe revolving within a sphere presses the surface of the sphere from Kepler's rule of the periodical times of the Planets being in sesquialterate [i.e., 3/2 power] proportion of their distances from the center of their Orbs, I deduced that the forces wch keep the Planets in their Orbs must [be] reciprocally as the squares of their distances from the centers about wch they revolve; & thereby compared the force requisite to keep the Moon in her Orb with the force of gravity at the surface of the earth, and found them answer [agree] pretty nearly."[5]

Note: For major qualifications of Newton's last statement, including evidence that in 1666 Newton viewed planetary motions

---

[2]This passage is from Newton's unsent letter of 1718 to Pierre Des Maizeaux as given in D. T. Whiteside, "Newton's Marvellous Year: 1666 and All That," *Notes and Records of the Royal Society, 21* (1966), 32. The Binomial Theorem provides a method for expanding and calculating binomials. The theorem states: $(x+y)^n = x^n + nx^{n-1}y + \frac{n(n-1)}{2!}x^{n-2}y^2 + \frac{n(n-1)(n-2)}{3!}x^{n-3}y^3 + \ldots + y^n$. For an example of the use of this theorem, see the derivation of the equation $E = mc^2$ in the chapter on Einstein and relativity.

[3]As quoted in Whiteside, "Newton's Marvellous Year," p. 32.

[4]As quoted in Whiteside, "Newton's Marvellous Year," p. 32.

[5]As quoted in Whiteside, "Newton's Marvellous Year," p. 32.

largely in the context of the Cartesian vortex cosmology, see the materials that follow.

1666	September 1: Beginning of the Great Fire in London.
1667	Returns to Cambridge; chosen a Minor Fellow of Trinity College; in March, 1668, a Major Fellow.
1668	July 7: Receives his M.A. degree.
1669	Early in 1669 contructs the first **reflecting telescope** ever built.
1669	August: Purchases some chemical equipment and also *Theatrum chemicum,* which consists of a collection of alchemical treatises.
1669	October 29: Succeeds Isaac Barrow as **Lucasian Professor of Mathematics**.
1670	Delivers the first of his optical lectures.
1671	December: Sends his reflecting telescope to the Royal Society.
1672	Elected (Jan. 11) to the Royal Society and publishes a paper in the *Philosophical Transactions of the Royal Society* announcing his discovery of the **heterogeneity of white light**, i.e., that white light, such as that coming from the Sun, consists of rays of all the spectral colors. Thus a ray of sunlight passed through a prism at certain angles will produce, rather than a white image of the Sun, an elongated image showing all the spectral colors from red through violet. This paper leads to a controversy with Robert Hooke, Curator of Experiments for the Royal Society, who maintains that the prism produces the colors whereas Newton correctly maintains that all the colors are present in white light and that the prism acts simply to separate these colors.
ca. 1672	Begins serious study of theology and of biblical prophecies. Continues these studies until 1684. Adopts Arian views, that is, he denies the divinity of Christ.
1675	Newton sends his second major paper on light and color to the Royal Society. Meets Robert Boyle (1627–1691), a leading chemist of the period. Around 1675, Newton was making plans to leave Cambridge. At that time, Cambridge fellows were required to take ordination within seven years of accepting their fellowship. Newton could not do this because a period of intense theological reading had led him secretly to adopt Arian convictions. Nonetheless, he is able to remain at Cambridge because of a special Royal Dispensation (27 April), which freed the Lucasian professor from the requirement for ordination.
1676	Summer and Autumn: Corresponds with G. W. Leibniz regarding mathematics.

1679   Letter of February 28 to Robert Boyle on Natural Philosophy. Death (early June) of Newton's mother.

1679–80 **Correspondence with Robert Hooke.** This stimulates Newton to develop further his theories in mechanics; in particular, Newton finds that bodies moved by a force originating from a point (i.e., a central force) must sweep out equal areas in equal time, as Kepler's second law of planetary motion indicates. Newton is also able to show that bodies (e.g., planets) moving in an inverse square field of force should move in ellipses.

1681   **John Flamsteed**, director of Greenwich Royal Observatory, suggests that a **comet** that had become visible in November, 1680 and had disappeared near the Sun on 8 December 1680, is identical to a comet that had appeared on the opposite side of the Sun on 10 December 1680, remaining visible until March, 1681. Flamsteed's idea ran counter to the traditional view that comets move either very nearly or exactly in straight lines. Newton questions not only Flamsteed's claim that a single comet had been seen but also Flamsteed's rather dubious explanation of this event. Over the next few years, however, Newton comes to agree with Flamsteed's claim that only a single comet had been seen; moreover, Newton devises a method for applying his mechanical ideas to cometary orbits.

1684   January: **Edmond Halley**, **Robert Hooke**, and **Sir Christopher Wren** meet and share the conclusions each had reached that gravitational force must vary inversely as the distance squared. Wren offers a prize to the person who can prove the correctness of this claim.

         August: Visit from **Edmond Halley** leads Newton to draw together his ideas in mechanics into his *De motu corporum in gyrum* (*On the Motion of Bodies in an Orbit*), which in November, Newton sends to Halley in London. It forms a sort of zeroth draft of the *Principia*. Appearance of Leibniz's first publication on his newly developed calculus.

1685   February: Death of King Charles II; succeeded by King James II. November: Newton finishes his *De motu sphaericorum corporum in fluidis* (*On the Motion of Spherical Bodies in Fluids*), a treatise over ten times longer than his nine page *De motu corporum in gyrum* of the previous November. It is the chief intermediary manuscript between his *De motu corporum in gyrum* of 1684 and his *Principia* of 1687.

1686–87	Newton completes the manuscript of his *Principia mathematica* and sends it to the Royal Society after an intense period (about 18 months) of writing.
1687	Publishes (5 July) his masterwork: **Philosophiae naturalis principia mathematica**.
1688	Nov./Dec.: **Glorious Revolution**, in which Protestant William and Mary replace the Catholic James II as rulers of England.
1689	Elected to **parliament** as a representative of Cambridge University; elected again in 1701. Meets John Locke (1632–1704), one of the most prominent English philosophers.
1692–3	Newton composes four letters to the theologian Richard Bentley; in these letters, Newton presents his views on the theological implications of his *Principia*.
1693	Spring and summer: Composes *Praxis*, his most significant alchemical treatise.
1693	Late summer: Newton suffers a severe mental breakdown.
1695	Death of Christiaan Huygens.
1696	March 19: Newton named Warden of the Mint; moves to London.
1699	December 25: appointed **Master of the Mint**, which position he holds until his death.
1701	Newton resigns (10 Dec.) the Lucasian professorship; William Whiston chosen to succeed him. Newton also at this time resigns his fellowship at Trinity College.
1702	Death of King William III. Accession of Queen Anne.
1703	March 3: death of Robert Hooke. November 30: Newton elected **President of the Royal Society**; remains President until his death.
1704	Newton publishes his **Opticks** with 16 queries at the end of it; Latin edition appears in 1706 with 7 new queries; 2nd English edition in 1717 with 8 new queries (these appear as #17–24, the 7 queries added in 1706 becoming #25–31; 2nd Latin ed. in 1719; 3rd English ed. in 1721. Two mathematical treatises were appended to the *Opticks*.
1705	April 16: knighted by Queen Anne. Henceforth, Sir Isaac Newton.
1707	Whiston publishes Newton's algebraic lectures, *Arithmetica universalis*.
1711	Newton publishes his *De analysi per aequationes numero terminorum infinitas* (*On Analysis by Means of Equations with an*

Infinite Number of Terms), which is Newton's first published exposition of his calculus, although various aspects of his calculus had appeared in his *Principia*.

1712   Royal Society sets up committee to examine a dispute between Leibniz and Newton concerning priority in discovering the calculus.

1713   Publication (30 June) of second edition of *Principia*, edited by Roger Cotes; third appeared in 1726, edited by Henry Pemberton. Publication of *Commercium epistolicum*, the report of the committee set up by the Royal Society to decide the dispute between Leibniz and Newton on priority regarding the discovery of the calculus. The committee sides with Newton, but was itself heavily influenced both in its composition and conclusions by Newton.

1714   Death of Queen Anne and accession of King George I.

1715   November: Beginning of the famous **Clarke-Leibniz correspondence** in which Samuel Clarke, writing for Newton, debates with Leibniz on such topics as whether the universe is a machine and whether it is proper to speak of absolute space and time.

1716   Death of Leibniz.

1717   First publication of the Clarke-Leibniz correspondence (see 1715).

1727   March 20: **Death of Newton**, who left an estate valued at £30,000; buried in Westminster Abbey.

1728   Publication of Newton's *Chronology of Ancient Kingdoms Amended*.

1729   Publication of Newton's *Lectiones opticae* (*Optical Lectures*).

1731   Unveiling of the monument to Newton in Westminster Abbey.

1733   Publication of Newton's *Observations upon the Prophecies of Daniel and the Apocalypse of St John*.

Introduction

Alexander Pope's famous couplet

> Nature, and Nature's Laws lay hid in Night:
> God said, *Let Newton be!* and All was *Light,*[6]

is no doubt overstated. Nonetheless, Newton's accomplishments were among the most remarkable ever attained. In discussing Newton's thought, it is well to begin both broadly and chronologically, focusing first on some of the ideas Newton explored during the 1665–1666 period, when Newton, who had but shortly before completed his B.A. degree, was forced to return to his home because of the plague that had closed Cambridge University. Beginning the story at this point has the advantage of providing a preliminary overview of some of Newton's most fundamental insights, which he presented in greater detail and complexity in 1687 in his *Mathematical Principles of Natural Philosophy*. The central question that the following materials attempt to answer is: How and when did Newton discover his famous **law of universal gravitation**? This law is that any two bodies in the universe attract each other with a force that is proportional to the product of the masses of the two bodies and inversely proportional to the square of the distance between them.

It is significant to note that Newton does not anywhere in his *Principia* express his law of gravitation in the form of an equation. We can, however, do this in the following manner. Let F represent the force of gravitational attraction, M_1 the mass of the first body, M_2 the mass of the second body, and D the distance between the bodies.

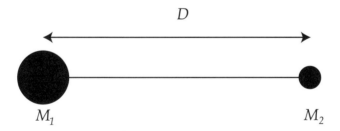

The equation is: $F \propto \dfrac{M_1 M_2}{D^2}$.

This proportionality has a number of features that are worth noting separately. The most obvious of these is the inverse-square force law; indeed,

---

[6] Alexander Pope, *Minor Poems,* ed. by Norman Ault, completed by John Butt (Yale University Press, 1954), p. 317; see also Derek Gjertsen, *The Newton Handbook* (Routledge and Kegan Paul, 1986), p. 439.

people sometimes jump to the conclusion that this is all there is to universal gravitation. The masses, however, are also involved, and not just the mass of the body being attracted, but also that of the attracting body. The force seems to be generated jointly by both of them. Finally, what does not appear in the algebra but is of crucial importance is that the attraction is between all the individual particles of the two bodies, and not between the bodies as wholes. Newton recognized that he had to demonstrate each of these points in order to establish universal gravitation.

Background: The Period before Newton

Let us look first at some of the resources available in the 1660s to persons interested in physical science.

1. By this time the **Copernican system** was widely accepted. Evidence indicates that Newton probably accepted it during his undergraduate days, if not earlier.

2. By 1621, **Johannes Kepler** had formulated what are now called the **three Keplerian laws of planetary motion**. These are:

   - that every planet moves in an ellipse with the sun at one focus of the ellipse;
   - that the planets sweep out equal areas in equal times; and
   - if the distance D of any planet from the sun is cubed and this quantity is divided by the square of the period T of the planet, then this quantity, $\frac{T^2}{D^3}$, equals a constant, which is the same for any planet.

   It is important to stress that Kepler had not himself presented these laws as a set of three laws; rather they are scattered in his writings. Moreover, they attracted only limited attention for a number of decades. One indication of their marginal status in astronomy is that Galileo never referred to them. Even by the 1660s, they, and especially the second law, remained somewhat in question. This may have been due to their status as for the most part empirical generalizations, to the fact that the first and second laws were developed in a book focused on only the planet Mars, and/or the hesitancy that astronomers had for the physical ideas that in Kepler's mind supported them. By the mid-1660s, Newton had become aware of at least the first and third Keplerian laws from reading Thomas Streete's *Astronomia Carolina* (1661), which did not mention the second law. By 1676 or earlier, he had also come to know of Kepler's second law.

3. By the end of the 1630s, **Galileo** had published his crucial contributions to terrestrial mechanics. Among the results Galileo had obtained was evidence that bodies on Earth fall at a rate that is independent of their weight and that is proportional to the time of their fall squared. Galileo had also attained an early form of the law of inertia and used it in analyzing the motions of bodies, for example, projectiles, which he showed follow a parabolic path. It is important to realize that Galileo's treatment of terrestrial motion was **kinematic**; that is, he had supplied a descriptive rather than a dynamic or an explanatory account of the motions of the bodies treated.

4. **René Descartes**, especially in his *Principia philosophiae* (1644), had formulated a system of the universe that purported to be explanatory. In particular, Descartes claimed that all space is filled with an **aether** and that the motions of the planets result from **vortices** in this aether. Part of Descartes's system was his claim that space cannot be empty; in other words, he denied that a void can exist. Another feature of his system was the claim that all forces must be explained by contact; there cannot be forces acting across empty space. It is important to realize that Descartes's explanations were largely in qualitative terms; nonetheless, his system seemed capable of detailed quantitative elaboration and was quite attractive to a number of his contemporaries. Two of its keystones were **the law of inertia** and Descartes's claim that **the quantity of motion in the universe is constant**. The chief way in which Descartes's formulation of the law of inertia differs from that of Galileo is that for Galileo inertia is circular; the great Italian physicist explained planetary motions by asserting that inertia acts in a circular manner. Regarding quantity of motion, Descartes asserted that God in the beginning had created a certain amount of motion in the universe and that this quantity remains unchanged. He tended to identify this quantity of motion as proportional to both the speed of the body and its magnitude. This quantity is what was symbolized by MV (mass times velocity) after the Newtonian concept of mass became available. Another feature of the Cartesian system was that Descartes, like Kepler before him, believed that **comets move in straight or nearly straight lines**. This belief caused problems for Newton and others who gradually became interested in explanations of celestial motions in terms of some sort of gravitational attraction. Newton read Descartes's *Principles of Philosophy* in late 1664 and was drawn to various features of it. This is a surprising fact to those who know how vigorously Newton attacked Cartesian physics in his *Principia*.

5. The **Epicurean System**, which was championed in the mid-seventeenth century by such figures as Pierre Gassendi and Walter Charleton, was quite different from the Cartesian system. The atomists believed that the universe is composed of atoms and the void and that explanations of physical phenomena must proceed in terms of these entities. In late 1663 or early 1664, Newton read Walter Charleton's *Physiologia Epicuro-Gassendo-Charltoniana* (1654), finding it attractive in a number of ways.

The Prehistory of Newton's *Principia*[7]

Newton during the 1660s

When Newton entered the University of Cambridge in 1661, the level of learning there was not high nor was much demanded of undergraduates. Newton's friend Abraham DeMoivre recounted how Newton began to become deeply involved with mathematics and science:

---

[7]The birthing of the most brilliant book ever published in physics is, naturally, an attractive subject for historians of science. Among the most accessible accounts of the development of Newton's thought leading to his *Principia* are various essays by Curtis Wilson; see especially Wilson's "Newton's Path to the *Principia* ," *Great Ideas Today 1985* (Encyclopædia Britannica, 1985), 179–229, and for more detail his "The Newtonian Achievement in Astronomy," *Planetary Astronomy from the Renaissance to the Rise of Astrophysics: Part A: Tycho Brahe to Newton*, ed. by R. Taton and C. Wilson (Cambridge University Press, 1989), 233–274. I have also profited from reading the appropriate sections of Richard Westfall's unsurpassed biography of Newton: *Never at Rest: A Biography of Isaac Newton* (Cambridge University Press, 1980). A very perceptive and detailed account is provided in chapters 10 and 11 of Julian B. Barbour, *Absolute or Relative Motion? A Study from a Machian Point of View of the Discovery and the Structure of Dynamical Theories*, vol. I: *The Discovery of Dynamics* (Cambridge University Press, 1989). Important insights can also be derived from J. Bruce Brackenridge, *The Key to Newton's Dynamics: The Kepler Problem and the* Principia (University of California Press, 1995). Much valuable information about this story is available in the writings of I. Bernard Cohen, especially Cohen's "Guide to Newton's *Principia*," which introduces (pp. 1–370) possibly the most important of all Cohen's contributions to Newtonian scholarship: Isaac Newton, *The Principia: Mathematical Principles of Natural Philosophy*, trans. I. B. Cohen and Anne Whitman (University of California Press, 1999). Another leading expert on Newton's mathematical and physical thought is Derek T. Whiteside; see his "The Prehistory of the *Principia* from 1664 to 1686," *Notes and Records of the Royal Society, 25* (1991), 11–61. Clarity, conciseness, and insight are all in evidence in the account of the *Principia* in A. Rupert Hall's biography of Newton: *Isaac Newton: Adventurer in Thought* (Cambridge University Press, 1996), esp. pp. 53–64, 162–171, and 202–224. Many of the key manuscript sources for the *Principia* are presented in John Herivel's *The Background to Newton's* Principia. *A Study of Newton's Dynamical Researches in the Years 1664–84* (Clarendon Press, 1965). Other valuable studies are listed in the bibliography at the end of this volume.

In [16]63 [Newton]. . . bought a book of Astrology, out of a curiosity to see what there was in it. Read in it till he came to a figure of the heavens which he could not understand for want of being acquainted with Trigonometry.

Bought a book of Trigonometry, but was not able to understand the Demonstrations.

Got Euclid to fit himself for understanding the ground of Trigonometry.

Read only the titles of the propositions, which he found so easy to understand that he wondered how any body would amuse themselves to write any demonstrations of them. Began to change his mind when he read that Parallelograms upon the same base & between the same Parallels are equal, & that other proposition that in a right angle Triangle the square of the Hypothenuse is equal to the squares of the two other sides.

Began again to read Euclid with more attention than he had done before & went through it.[8]

Contact with Euclid set Newton going in mathematics. He proceeded to read other mathematical books, including treatises by William Oughtred and by John Wallis, the Savilian Professor of Geometry at Oxford, as well as Descartes's *La géométrie* (1637), which is the first exposition of analytical geometry. By 1666, Newton had advanced to the point that he had invented the fundamental notions of differential and integral calculus. This discovery, if published, would have established the young and unknown Newton as the most creative mathematician in Europe. No one, however, was aware of what he had done. It would be interesting to pursue his mathematical work in more detail, but our chief concern at present is with Newton's contributions to mechanics.

Newton's friend William Stukeley, in a biography of the great scientist that he published in 1754, recorded that in 1726 while he was having tea with Newton under some apple trees, Newton stated that "he [Newton] was just in the same situation, as when formerly, the notion of gravitation came to his mind. It was occasion'd by the fall of an apple, as he sat in a contemplative mood."[9] What is supposed to have happened is that Newton, seeing the apple fall and realizing that in a sense the Moon also falls toward the Earth, concluded that the Earth exerts a gravitational attraction on both the Moon and the apple, which attraction acts analogously in both cases. From this, it is also repeatedly alleged, Newton went on to conclude that gravitational

---

[8] As quoted in *The Mathematical Papers of Isaac Newton*, ed. by D. T. Whiteside, vol. 1 (Cambridge University Press, 1967), pp. 5–6.

[9] As quoted with commentary in Gjertsen, *Newton Handbook*, p. 29. For the original, see William Stukeley, *Memoirs of Sir Isaac Newton's Life*, ed. A. Hastings White (Taylor and Francis, 1936), 19–20.

attraction is universal, extending to all material bodies. In short, this story is interpreted so as to provide a basis for the claim that in the mid-1660s, Newton formulated his famous theory of universal gravitation. Extensive research in Newton's manuscripts over the last five decades has revealed a number of reasons for rejecting this claim, despite Newton's statement and despite the widespread acceptance for many years of it as well as other comparable statements made by Newton. In particular, it is now evident that although Newton in the 1660s did attain some notions that were eventually useful to him in formulating his theory of universal gravitation, he did not in fact attain that theory until the 1680s. Rather it is the case that throughout the late 1660s and the 1670s, Newton was still searching for an explanation of motion in terms of an aether and vortices in that aether.

In analyzing how Newton derived his theory of universal gravitation, let us consider a statement made by Newton in a 1718 letter to Pierre Des Maizeaux referring to what Newton achieved in the period 1665–1666:

> [In 1666] I began to think of gravity extending to y^e [the] orb of the Moon & having found out how to estimate the force with w^{ch} [a] globe revolving within a sphere presses the surface of the sphere from Kepler[']s rule of the periodical times of the Planets being in sesquialterate [i.e., 3/2 power] proportion of their distances from the center of their Orbs, I deduced that the forces wch keep the Planets in their Orbs must [be] reciprocally as the squares of their distances from the centres about wch they revolve; & thereby compared the force requisite to keep the Moon in her Orb with the force of gravity at the surface of the earth, and found them answer [agree] pretty nearly.[10]

In this passage Newton seems to be saying that he had not only the idea of gravity in 1666, but also the idea of the inverse square law of gravitational attraction and that he had found a quasi-verification for the inverse square law for the case of the Moon.

Study of Newton's manuscripts shows that around early 1666, he began to think seriously and deeply about mechanics, doing this primarily in the context of the Cartesian system. He first took up the problems of collision theory, before long attaining results superior to those achieved by Descartes and comparable to the results reached by Huygens. Newton also took up the issue of circular motion, conceiving it initially in a Cartesian context. In particular, Newton conceived of circular motion as a motion explicable partly in terms of the law of inertia and partly by what Descartes had called a *conatus recedendi a centro* ("endeavor to recede from the center"). Descartes had conceived of this force in analogy with a stone pushing outwards on a sling as it is whirled around. Thus the Moon would experience both a tendency

---

[10] As quoted in Whiteside, "Newton's Marvellous Year" (footnote 2), p. 32.

to move in a straight line and also to move away from the center of its orbit. Newton's notion that the Moon would experience a force away from the center of its orbit raised the question: what would prevent it from moving in that direction? Part of the answer, he realized, lay in the tendency of bodies to move in straight lines. Nonetheless, Newton felt constrained to explain the Moon's motion by postulating an aether medium that would flow rapidly toward the center of the Earth, which aetherial particles would resist the centrifugal force on the Moon. This was a different interpretation from that offered by Descartes, but very much in the Cartesian tradition in that it assumed that aetherial material pervaded all space and motion could be explained only by contact forces, not by forces acting at a distance. Such was the physical picture that Newton held to until around 1680.

The mathematical approach used by Newton at this time for analyzing constant circular motion was complementary. By means of assumptions about the cumulative effect of the centrifugal force and by treating the motion of a ball rolling around the sides of a cylinder as if it were in fact moving along the sides of infinitely sided polygon inscribed in the cylinder, Newton succeeded in deriving in 1666 the law of centripetal force for constant circular motion: $F \propto \frac{v^2}{r}$, where v is the speed of the ball and r is the radius of the cylinder. On the assumption that force (F) and acceleration (a) are proportional, this also gives the law of centripetal acceleration: $a = \frac{v^2}{r}$. Later in the 1660s, Newton worked out another proof of the law of centripetal acceleration, using a quite different method of conceptualizing the problem, but deriving the same result. The law of centripetal acceleration was also formulated around this time by Christiaan Huygens, the great Dutch physicist, who, to Newton's distress, published it first in his 1673 treatise on pendulum clocks (*Horologium oscillatorium*).

Application of the Law of Centripetal Acceleration to the Moon

Let us use this law to compare the rate of acceleration for bodies on the Earth with the rate at which the Moon accelerates toward the Earth. In making this calculation, we shall use data now available, rather than the less accurate data available when Newton made a comparable computation.

Assume that the Moon moves in a nearly circular orbit of radius 239,000 miles and that it completes one revolution each 656 hours. Using our formula, we can calculate the rate at which it accelerates toward the Earth. To compute the speed of the Moon's motion, take the circumference of the circle of its orbit and divide this by its period.

$$v = \frac{2\pi \cdot 239,000 \text{ miles}}{656 \text{ hours}} = 2,289 \text{ mph}$$

By our equation $a = \frac{v^2}{r}$, we have

$$a = \frac{2289^2}{2.39 \cdot 10^5} = 21.9 \frac{\text{miles}}{\text{hr}^2}.$$

Converting this to feet and seconds, we have that the rate of the Moon's acceleration toward the Earth is $0.0089 \frac{\text{ft}}{\text{sec}^2}$.

Let us now compare this value to the rate of acceleration of bodies on the surface of the Earth; as is well known, the acceleration for such bodies is $32.2 \frac{\text{ft}}{\text{sec}^2}$. If we divide 32.2 by 0.0089, we get approximately $3,600$. What is noteworthy is that 3600 is the square of 60, which is the number of Earth radii that the Moon is distant from the Earth. In other words, it appears that the rate at which an apple falls to the Earth is to the rate at which the Moon falls toward the Earth inversely as the square of their distances from the center of the Earth.

How is this striking fact to be interpreted? Newton's fully formulated mechanics as given in his *Principia* provides an explanation. It is that the Earth is attracting both the Moon and the terrestrial body and that the attractive force exerted by the Earth on each body is inversely proportional to the square of the distance of each from the Earth's center. Interpreted in this way, i.e., in terms of the law of gravitation, this numerical result seems a strong confirmation of the law of gravitation. This interpretation would support Newton's statement in his 1718 letter to Des Maizeaux (cited earlier) about having made a calculation of the force the Moon exerts on the Earth.

Recent research, based on Newton's manuscripts, shows, however, that this interpretation is not correct; in particular, it is now clear that Newton in the period around 1666 was not thinking in terms of *universal* gravitation; rather he was primarily concerned with comparing rates of acceleration. In addition, he was, following Descartes, thinking in terms of there being a **centrifugal** (away from the center) force on the Moon. Moreover, Newton was at this time opposed to the idea of a universal gravitational force, which had some currency in the mid-seventeenth century, on the grounds that the idea of such a force seemed to be contradicted by the widespread belief that comets move in straight lines. In addition, the pattern evident from research on Newton's optical writings also applies to Newton's approach to mechanics. As A. Rupert Hall documents in his book on Newton's optical research, there was an "interval of roughly thirty years (1680–1710) in which Newton eschewed aetherial hypotheses in favor of force-mechanics...."[11] In other words, both before and after that thirty year period, Newton's preferred method for explaining physical phenomena was in terms of an aetherial fluid filling all space. Moreover, in his biography of Newton, Hall remarks that

---

[11] *All Was Light: An Introduction to Newton's* Opticks (Clarendon, 1993), p. 118.

"We know neither when nor why Newton threw away vortices and impacts [i.e., Cartesian physics] in favour of a pragmatic, mathematical mechanics of forces first sketched in *De Motu*. We can only record that he embarked on this astonishing new world as a result of a visit paid to him in Cambridge in August 1684 by Edmond Halley....[12]

The Relationships among the Inverse Square Law, Kepler's Third Law, and the Law of Centripetal Acceleration

In his 1718 letter to Des Maizeaux, Newton mentioned a calculation involving Kepler's third law. Around 1669, Newton did make such a calculation, which, at least when seen in hindsight, provided support for a law of universal gravitation. Let us examine one form of the calculation.

Recall that Kepler had shown that for any planet, $\frac{D^3}{T^2} = C$, where D is the distance of the planet from the Sun, T is the planet's sidereal period,[13] and C is a constant, which is the same irrespective of what planet is considered. Let us combine this astronomical formula with our geometrical formula that $a = \frac{v^2}{r}$. This is possible if we assume that the planets move nearly in circles and if we assume that each has a uniform acceleration towards the Sun. Since $D = r$, we have from Kepler that $T^2 = \frac{D^3}{C} = \frac{r^3}{C}$. We also know by the definition of velocity that for a planet moving at a constant speed v on a circular orbit of radius r that $v = \frac{2\pi r}{T}$. Let us now substitute our value for T^2 into the formula for centripetal acceleration. We then have:

$$a = \frac{v^2}{r} = \frac{(2\pi r)^2}{T^2} \cdot \frac{1}{r} = \frac{(2\pi)^2 r^2}{r^3/C} \cdot \frac{1}{r} = \frac{C(2\pi)^2 \cdot 1}{r^2}.$$

Because $C(2\pi)^2$ is a constant, we see that this result agrees with the notion that the centripetal acceleration toward the Sun of every planet is inversely as the square of its distance from the Sun. If force and acceleration are proportional, then the gravitational force due to the Sun should decrease in inverse proportion to the distance squared. This does not prove that there is a universal gravitational force that obeys the inverse square law, but it does provide a hint that would agree with such a notion. Newton was no doubt interested in this result, but it was not enough to lead him in the late 1660s to infer that an inverse square law governs planetary motion. A version of this proof appears in Book I of Newton's *Principia* as Corollary 6 to Proposition 4: "*If the periodic times be in the sesquiplicate ratio of the radii, and therefore the velocities inversely in the subduplicate ratio of the radii, the centripetal*

---

[12] Hall, *Isaac Newton*, p. 207.
[13] The sidereal period of the planet is the time it takes to pass around the Sun and realign with some star.

forces will be inversely as the squares of the radii, and conversely."[14] This statement if put into modern algebraic parlance would read: If the periodic times be proportional to the 3/2th power of the radii, and therefore the velocities be inversely as the square roots of the radii, the centripetal forces will be inversely as the squares of the radii, and conversely.

Summary

Drawing this together, we see that Newton in the 1660s had begun to investigate how celestial bodies move, that he had attained the powerful law of centripetal acceleration, that he had found evidence that could be made to agree more or less with an inverse square law of attraction, but that as yet he was still searching for some form of explanation in terms of aetherial vortices for the movement of the planets. His atomism made him dissatisfied with Descartes's analysis, but as yet he had no adequate theory of his own. All this indicates that as remarkable as Newton's *annus mirabilis* was, he still had a long way to go before securing the system he expounded in 1687 in his *Principia*.

Newton from 1670 to 1680, Especially His Correspondence with Robert Hooke

In pursuing how Newton attained the law of universal gravitation, we shall now turn to his contacts with the ideas of two other scientists, the Italian **Giovanni Alphonso Borelli** (1608–1679) and Newton's fellow Englishman **Robert Hooke** (1635–1703). We shall see that each of them possessed some notion of gravitational attraction. In 1666, Borelli published his *Theoricae mediceorum planetarum ex causis physicis deductae* (*Theory of the Medicean Planets Deduced from Physical Causes*), which book Newton read in the 1670s. In that volume, Borelli attempted to explain the motions of the planets and their moons by describing their motions as the result of a centrifugal tendency for bodies to move away from the Sun combined with a centripetal tendency for them to be drawn toward the Sun, the latter being a force originating in the Sun. Thus Borelli's book contained a primitive notion of a gravitational force extending throughout the solar system. Although Borelli accepted the law of inertia, he failed to make proper use of it in explaining these motions, as is now known to be necessary.

---

[14] As translated in Dana Densmore, *Newton's* Principia: *The Central Argument*, with translations and illustrations by William H. Donahue, 3rd ed. (Green Lion Press, 2003), p. 164.

Robert Hooke did rather better. Let us look at some ideas expressed in 1674 by Hooke in his *An Attempt to Prove the Motion of the Earth*, which Newton probably read before the end of the 1670s. In that book, Hooke laid out three suppositions; his first supposition is that

> ...all celestial bodies whatsoever have an attraction or gravitating power towards their own centers, whereby they attract not only their own parts...but...also...all the other celestial bodies that are within the sphere of their activity; and consequently that not only the Sun and Moon have an influence upon the body and motion of the Earth, and the Earth upon them, but that Mercury and also Venus, Mars, Saturn, and Jupiter by their attractive powers, have a considerable influence upon its motion as in the same manner the corresponding attractive power of the Earth hath a considerable influence upon every one of their motions also.[15]

In short, Hooke's first supposition seems to be that there is a widespread force of attraction among the celestial bodies. His second supposition is simply the law of inertia. His third supposition is that "these attractive powers are so much the more powerful in operating, by how much the nearer the body wrought upon is to their own centers." Hooke then admitted that he did not know quantitatively what this attractive power is, but he promised to investigate it experimentally. Hooke's statements may seem very close to the law of gravitation; however, they fall short in a number of ways. First, he lacked a determination as to how the force of attraction varies with distance. Also his attraction, which he conceived of in analogy with magnetic attraction, was not actually universal. This is indicated both by his reference to "sphere of activity" and also by the fact that Hooke, believing that comets move in straight lines, suggested in a 1678 book titled *Cometa* that the nature of the matter of comets is such that gravitational attraction does not act upon them.

In 1679, Newton received a letter from Robert Hooke. Newton had previously come in contact with Hooke in regard to some optical writings of Newton. These contacts had made for strained relations between the two scientists. In fact, Newton always remained hostile to Hooke. The cluster of letters that they exchanged in 1679–1680 is very important because these letters helped Newton to a number of important recognitions or discoveries. For example, it appears that this contact with Hooke led Newton to see that the proper way to conceptualize the motion of the Moon and planets was as resulting from the interactions of the tendency of bodies to move in a straight

---

[15] As quoted in Curtis Wilson, "Newton's Path to the *Principia*" in *Great Ideas Today 1985* (Encyclopædia Britannica, 1985), p. 197.

line at constant speed and a tendency in the Moon or planet to be attracted toward the body at the center of its orbit.[16] Moreover, in pondering some problems of mechanics raised by Hooke, Newton was led to a proof that bodies moving under the influence of a force exerted by some body (that is, in what is called a "central field of force"), must sweep out equal areas in equal times. This is the first proposition of his *Principia*. What this proposition states is that if a central body exerts a force, no matter how quantitatively that force varies with distance, then bodies moving under the influence of that force, must sweep out equal areas in equal times. One significance of this was that Newton had found a method of proving that Kepler's second law must be the case. Newton also discovered evidence that the attraction between celestial bodies varies as the inverse square. Hooke in the correspondence had expressed the conviction that this is the case, but Hooke lacked proof for this claim. What Newton did was to show (at least, so he was later to maintain—see below) that if an attraction exists and is proportional to the inverse square of the distance, then planets must move in ellipses. Newton did not communicate his proof at this time either to Hooke or to anyone else. It is significant to note that although Hooke deserves credit for having an idea of an inverse square law of attraction, the great achievement was to link this supposition mathematically with such empirical evidence as Kepler's first law. And this Newton alone achieved.

Newton, Flamsteed, and the Comet of 1680–1681

Another step in Newton's thought came in 1680–1681. In November, 1680, a comet had appeared moving toward the Sun. Astronomers last saw the comet on 8 December 1680, at which time it was very near the Sun. On 10 December 1680, another comet (or so it was thought) became visible on the opposite side of the Sun. That comet remained visible until March, 1681. Among those observing it was the English astronomer John Flamsteed, who suggested that these were not two separate comets but rather a single comet, the orbit of which had been altered by the action of the Sun. Flamsteed worked out a theory as to how this process could have taken place. Newton found a host of objections, most of them legitimate, to Flamsteed's theory. Moreover, at this time, Newton remained unconvinced that these two cometary apparitions were due to a single comet; in fact, in correspondence with Flamsteed he argued against this position. Over the next few years, however, Newton changed his mind. Not only did he conclude that

---

[16]Curtis Wilson, "The Newtonian Achievement in Astronomy," *Planetary Astronomy from the Renaissance to the Rise of Astrophysics: Part A: Tycho Brahe to Newton*, ed. by R. Taton and C. Wilson (Cambridge University Press, 1989), 233–274:240.

a single comet had been seen, but also that his gravitational ideas could be applied to comets as well as to planets. And he worked out mathematical methods for doing this.

Newton from 1684 to 1687

The idea that gravitational attraction exists and that it varies as the inverse square of the distance was "in the air" in the early 1680s. This is evident from a conversation that took place in January of 1684. Participating in this conversation were Hooke, **Edmond Halley** (1656–1742), and an astronomer turned architect, **Sir Christopher Wren** (1632–1723). All three had hit upon the idea that the force varies inversely as the distance squared. One of Halley's letters records what was discussed in that very important conversation.

> ...I met with Sir Christopher Wren and Mr. Hooke, and falling in discourse about it [the inverse square law], Mr. Hooke affirmed that upon that principle all the laws of celestial motion were to be demonstrated, and that he himself had done it; I declared the ill success of my attempts; and Sir Christopher, to encourage the Inquiry said, that he would give Mr. Hooke, or me, two months time, to bring him a convincing demonstration thereof; and besides the honour, he of us, that did it, should have from him a present of a book of 40 shillings.[17]

Neither Hooke nor Halley succeeded in this challenge, which is one of a number of reasons for concluding that, although Hooke was ingenious, he had not as yet solved the problem, despite his claim to the contrary.

In August, 1684, Halley happened to go to Cambridge, where he visited Newton, who had as yet published none of his ideas on mechanics, although during the 1670s he had attracted much deserved attention because of his brilliant writings on the nature of light. Halley asked Newton what would be the orbit of a planet moving under a gravitational attraction decreasing inversely as the distance squared. When Newton immediately replied that it would be an ellipse, Halley asked him how he knew this would be the case. Newton stated that he had proven it, but being unable to find his proof, agreed to send Halley a copy of it. Newton's response, at least if it can be taken at face value, is rather striking: he was in effect stating that he had mislaid the most important calculation of the century! In November, 1684, however, Halley received from Newton a nine page manuscript, *De motu corporum in gyrum* (*On the Motion of Bodies in an Orbit*), in which Newton derived Kepler's second law, and proved that the other Keplerian laws are

---

[17] Letter of 29 June 1686 from Edmond Halley to Isaac Newton as given in *The Correspondence of Isaac Newton*, ed. by H. W. Turnbull, vol. 1 (Cambridge University Press, 1960), p. 442. Note: I have slightly modernized the spellings in this letter.

consistent with the inverse square law. Halley realized the importance of Newton's results and encouraged him to publish them. Newton embarked on an intense period of activity, which culminated with the delivery of the manuscript of the *Principia* to the Royal Society. Published in July, 1687, it eventually came to be recognized as the most important book ever published in the history of the physical sciences.

In summary, the long accepted and widely deseminated story of how Newton in a single year in the 1660s discovered the fundamental insights that more than two decades later he presented in his epoch making *Principia* turns out to be very far from the truth. It was only in the period from 1679 to 1684 that Newton freed himself from the Cartesian cosmology and proceeded to attain his central force theorem, his law of universal gravitation, and the overall system that he so brilliantly embodied in his *Principia*. Moreover, in this process, Newton's efforts drew significantly on the writings and ideas of an array of authors, including Descartes, Hooke, and Flamsteed, to whom, for various reasons, Newton gave very limited credit. In fact, the claims Newton made to Stukeley and Des Maizeaux (cited earlier in this chapter) about when and how he discovered universal gravitation are now seen as motivated in substantial part by his concern to establish priority regarding his discovery by dating its origin back a decade and half before he attained it.

Newton's *Principia*[18]

Newton's *Principia* is a difficult book to read; not only is the subject matter complex, but also its presentations are in many ways products of Newton's time. Modern day readers encounter him criticizing arguments, formulating proofs, and making distinctions that sometimes seem to have little present

---

[18]A number of studies have been published of the nature, content, and structure of Newton's *Principia*. The most accessible of these is: Dana Densmore, *Newton's* Principia: *The Central Argument*, with translations and illustrations by William H. Donahue, 3rd ed. (Green Lion Press, 2003). See also Densmore's *Selections from Newton's* Principia: *A Science Classic Module for Humanistic Studies* (Green Lion Press, 2004) and Densmore's "At What Point in *Principia* Does Newton Believe Universal Gravitation Is Established?" in *Essays in Honor of Curtis Wilson* (MIT Press, forthcoming). Much valuable information is available in I. Bernard Cohen's "A Guide to Newton's *Principia*," which appears as pp. 1–370 in Isaac Newton, *The Principia: Mathematical Principles of Natural Philosophy*, trans. I. B. Cohen and Anne Whitman (University of California Press, 1999). See also J. Bruce Brackenridge, *The Key to Newton's Dynamics: The Kepler Problem and the* Principia (University of California Press, 1995) and Niccolò Guicciardini, *Reading the* Principia: *The Debate on Newton's Mathematical Methods for Natural Philosophy from 1687–1736* (Cambridge University Press, 1999). In the last years of his life, the Nobel Prize winning astrophysicist Subrahmanyan Chandrasekhar published his *Newton's* Principia *for the Common Reader* (Clarendon, 1995). Various older commentaries are also useful.

significance. One form this problem takes concerns the **mathematical methods** used in the *Principia*. It is frequently claimed that Newton decided to forego the use of his newly invented calculus, preferring instead to present the *Principia* in classical geometrical garb, even though, it is further assumed, he must have derived many of his results by means of the calculus. The latter assumption is not supported by examination of his manuscripts; one rarely if ever finds in them cases in which Newton derived a result by means of the calculus. Rather they appear to have been first formulated by means of the somewhat cumbrous Euclidean, Archimedean, and Apollonian geometrical methods in which they usually appear in the *Principia*. Nonetheless, it is significant that many of Newton's proofs in the *Principia* involve ideas and methods of the differential and integral calculus. Newton himself pointed this out when in the early eighteenth century he became involved in a priority dispute as to who had first discovered and published the main theorems of the calculus. Newton's use of calculus in the *Principia* was not in the form or formalism that later became common in the eighteenth century. Nonetheless, in a substantial number of cases, ideas and methods of the calculus appear in the *Principia*, increasing the difficulties for contemporary readers.

For these reasons and because of the length of the book, we shall study only its fundamental arguments and attempt from them to see its overall structure. This is possible and should provide a sound understanding of one of the greatest achievements of the human intellect. Moreover, it will help lay the background for an understanding of Einstein's theory of relativity.

Newton's *Principia* is divided into three books. The first book, which is highly mathematical, contains relatively few direct applications to our world. Early in Book I, Newton lays out his three laws of motion, but without providing extensive empirical evidence in support of them. From those laws he works out different possibilities of what force laws might be at work governing centripetal forces. For example, if we have bodies moving in eccentric circles, the force law would take one form, if in spirals, another, and if in ellipses, yet another form. In short, Book I is largely hypothetical and theoretical. Then in Book III, Newton presents empirical evidence showing that the inverse square law, especially when combined with other laws and propositions, does apply to our physical universe. Book II of the *Principia*, which focuses on the motions of bodies in resisting media, appears to be an anomaly. In fact, one of Newton's chief goals in it was to refute the vortex cosmology championed by Descartes. Descartes had argued that there are a number of possible mechanical systems, that the vortex system is one possible system of the world, and indeed that the vortex system represents our world. Newton had to show that the vortex system is impossible if he were to show that his possible world is the actual world. Consequently, he devoted much of Book II of the *Principia* to refuting Cartesian mechanics. Newton's debt to Descartes was quite large, but this is obscured by the vehemence with which he rejected the Frenchman's system. The law of inertia is but one of the truths Newton obtained from his 1664 reading of Descartes's *Principles of Philosophy*. We shall not be further concerned with Book II of the *Principia*.

In reading the selections from the *Principia*, it will prove helpful to attempt to formulate answers to the following questions:

1. For the most part, Galileo's mechanics was kinematical, that is, Galileo took as his task a description of motion, in particular, terrestrial motion. Newton's goal seems to have been broader, to have been dynamical; that is, he attempted to explain why bodies move as they do. In reading Newton, it is productive to ask: to what extent was Newton's approach truly dynamical and explanatory?

2. Newton claimed that his basic laws were induced from the phenomena. This would imply that metaphysical ideas did not enter into his system, although the system might itself have metaphysical implications. Is Newton's claim as to the inductive basis of his mechanics correct and is his system free from metaphysical assumptions?

3. To what extent was Newton indebted to his predecessors? One difficulty in answering this question is that in his *Principia* Newton mentioned his predecessors only relatively rarely and, in a number of these cases, e.g., that of Descartes, only as part of an effort to refute their claims. You will find, nonetheless, that in some cases you will be able to shed light on aspects of this question.

4. What were the theological implications of Newton's system, especially as seen by its author?

In reading Newton, an effective approach is to read the assigned passage with some care; then, after studying the commentary provided, return to the text, examining it in more detail.

Sir Isaac Newton

Mathematical Principles
of

Natural Philosophy
Third Edition
1726

Newton's Preface to the Reader[19]

The ancients, as Pappus wrote, made mechanics of the highest value in the investigation of natural matters, and more recent writers, having dismissed substantial forms and occult qualities, have made an approach to referring the phenomena of nature back to mathematical laws. It has accordingly seemed fitting in this treatise to develop mathematics insofar as it looks to philosophy. Now the ancients established two branches of mechanics: rational, which proceeds accurately by demonstrations; and practical. To practical mechanics all the manual arts look, and from here its name "mechanica" is borrowed. But since artisans are accustomed to work with little accuracy, it happens that mechanics as a whole is so distinguished from geometry, that whatever is accurate is referred to geometry, and whatever is less accurate, to mechanics. The errors, however, belong to the artisan, not the art. One who works less accurately is a more imperfect mechanic, and if any could work with perfect accuracy, this would be the most perfect mechanic of all. For the drawing of both straight lines and circles, upon which geometry is founded, belongs to mechanics. Geometry does not teach how to draw these lines, but requires [*postulat*] that they be drawn. For it requires that the beginner learn to draw them accurately before crossing the threshold of geometry, and then teaches how problems are solved by these operations. To draw straight lines and circles are problems, but not geometrical problems. The solution of these is required of mechanics, and once the solutions are found, their use is taught in geometry. And it is the glory of geometry that so much is accomplished with so few principles that are obtained elsewhere. Thus geometry is founded upon mechanical procedure, and

---

[19]Dana Densmore, *Newton's* Principia: *The Central Argument*, with translations and illustrations by William H. Donahue, 3rd ed. (Green Lion Press, 2003), pp. 3–4.

is nothing else but that part of universal mechanics that accurately sets forth and demonstrates the art of measuring. Further, since the manual arts are chiefly concerned with making bodies move, it happens that geometry is commonly related to magnitude, and mechanics to motion. In this sense, rational mechanics will be the science of the motions that result from any forces whatever, and of the forces that are required for any motions whatever, accurately set forth and demonstrated. This part of mechanics was developed into five powers by the ancients, looking to the manual arts, since they considered gravity (which is not a manual power) not otherwise than in the weights that were to be moved by those powers. We, however, are interested, not in the arts, but in philosophy, and write of powers that are not manual but natural, treating mainly those matters pertaining to gravity, levity, elastic force, the resistance of fluids, and forces of this kind, whether attractive or impulsive. And on that account we present these [writings] of ours as the mathematical principles of philosophy. For the whole difficulty of philosophy appears to turn upon this: that from the phenomena of motion we may investigate the forces of nature, and then from these forces we may demonstrate the rest of the phenomena. And to this end are aimed the general propositions to which we have given careful study in the first and second books. In the third book, on the other hand, we present an example of this procedure, in the unfolding of the system of the world. For there, from the celestial phenomena, using the propositions demonstrated mathematically in the preceding books, we derive the forces of gravity by which bodies tend to the sun and the individual planets. Then from the forces, using propositions that are also mathematical, we deduce the motions of the planets, of comets, of the moon, and of the sea. In just the same way it would be possible to derive the rest of the phenomena of nature from mechanical principles by the same manner of argument. For I am led by many reasons to strongly suspect that all of them can depend upon certain forces by which the particles of bodies, by causes not yet known, either are impelled towards each other mutually and cohere in regular shapes, or flee from one another and recede. These forces being unknown, philosophers have hitherto probed nature in vain. It is my hope, however, that the principles set forth here will shed some light either upon this manner of philosophizing, or upon some truer one.

In the remainder of his Preface, the chief point that Newton stresses is the important role that Edmond Halley played in encouraging Newton in the writing and publication of his *Principia*. Newton wrote brief prefaces for the second and third editions of the *Principia*, devoting these to noting the changes and additions made in these editions and (for the third edition) warmly thanking Henry Pemberton for his role in aiding Newton with the preparation of the third edition, which was the last published during Newton's lifetime.

Newton's Definitions[20]

Definition 1

The quantity of matter is the measure of the same arising from its density and magnitude conjointly.

Air of double density, in a space that is also doubled, is quadrupled; in a tripled [space], sextupled. The same is to be understood of snow and powdered substances condensed by compression or liquefaction. And the same account is given of all bodies which are condensed in various ways through various causes. In this I do not take account of the medium (if any) freely pervading the interstices of the parts. Further, in what follows, by the names "body" or "mass" I everywhere mean this quantity. It is apprehended through an individual body's weight. For it is found by experiments with pendulums carried out with the greatest accuracy to be proportional to weight, as will be shown hereafter.

Commentary on Definition 1

At first sight, it may seem that Newton's definition of mass makes no sense. Is not mass the same as weight? Not for Newton! Recall that in his system, the Earth attracts an apple with a certain force; this force manifests itself as weight. Suppose the apple were removed to sixty Earth radii away from the Earth, that is, to the Moon's orbit. Would it weigh the same, or, to put it another way, be attracted by the same force? In the Newtonian system, the force on it would be 1/3600 of the force on it were it located on the Earth's surface. From this, we see that **weight is relative** and that Newton needs to specify some property of bodies that does not vary with the location of the body. The concept he selected was **mass**. Scholars indicate that Newton arrived at the notion of mass only fairly late in the process of writing his *Principia*; in fact, it was only in early 1685 that Newton introduced the notion of mass into his manuscripts.[21] And nowhere in the *Principia* did he attempt to define a unit for mass.

---

[20] Densmore, *Central Argument* (footnote 19), p. 5.

[21] I. B. Cohen and R. S. Westfall (eds.), *Newton: A Norton Critical Edition* (W. W. Norton, 1995), p. 222.

Newton's distinction is preserved in modern day physics textbooks, where one immediately learns how to calculate the mass of a body from its weight, provided that its position (e.g., distance from the Earth's center) is given. In the light of historical perspective, we can see that one important feature of Newton's definition of mass is not mentioned in it. Newton assumed that the mass of a body does not depend upon its velocity, which is surely a natural assumption. This seemingly natural assumption, however, was challenged by **Albert Einstein**, Newton's great successor in theoretical mechanics.

Principia Example

One formula relating force and mass is $F = ma$. This formula when applied to weight stipulates that the weight of a body in a given region of space is equal to its mass times the acceleration it would experience in free fall. The unit of mass in the metric system is the **kilogram** and the unit of force or weight is the **newton**. The definition of a newton is the force that will impart to a 1 kilogram mass an acceleration of 1 $\frac{\text{meter}}{\text{sec}^2}$. Suppose a body has a mass of 10 kilograms. What will it weigh on Earth?

Solution: From knowing that bodies on Earth fall at 9.8 $\frac{\text{meters}}{\text{sec}^2}$ and that $F = ma$, we can conclude that $F = 10$ kilograms $\cdot\ 9.8\ \frac{\text{meters}}{\text{sec}^2}$. Hence it will weigh 98 newtons.

Problem

Suppose you weigh 800 newtons on Earth. Calculate your mass and from it determine your weight on Mars, where bodies fall at $3.77\ \frac{\text{meters}}{\text{sec}^2}$.

Solution: By $F = ma$, we have $800 = m \cdot 9.8$; hence $m = 81.6$ kilograms. Your mass is only 81.6 kilograms! Your weight on Mars will be $F = 81.6 \cdot 3.77 = 308$ newtons.

Newton's Definitions, continued[22]

Definition 2

The quantity of motion is the measure of the same arising from the velocity and the quantity of matter conjointly.

The motion of the whole is the sum of the motions in the individual parts, and therefore in a body twice as big, with an equal velocity, it is doubled, and with a doubled velocity it is quadrupled.

---

[22] Densmore, *Central Argument* (footnote 19), p. 8.

Commentary on Definition 2

Newton's definition of **quantity of motion** sets it equal to the mass of the body times its velocity, that is, mv. This quantity is now known as **momentum**. It is one of the most important conceptual categories in Newtonian mechanics. Its importance is suggested by the fact that it has a long and distinguished history. Jean Buridan, for example, in discussing the quantity that he called "impetus" described it as being measured both by the weight and by the speed of the body. Galileo also integrated this quantity into his mechanics, using it to understand various phenomena, for example, the tendency of a pendulum bob to return to the same height in its swing.

One reason scientists came to recognize the importance of the conceptual entity momentum is that momentum is a conserved quantity; for example, in collisions between hard bodies, the sum of their momenta remains the same before and after the collision. This facilitates the solution of an array of problems. In the background of Newton's definition of quantity of motion is the claim made by **Descartes** that God in the beginning created in the universe a certain definite and unchanging quantity not only of matter but of motion as well. Problems arose as to how this quantity of motion should be defined. Newton favored mv, whereas **Leibniz**, in a paper published in 1686,[23] urged that the conserved quantity of motion is not mv, but mv^2. The latter conceptual category was not used by Newton in his mechanics, but was employed during the eighteenth century by many continental physicists, who called it "**vis viva**" or "living force." This quantity later came to be designated, as it is now, as **kinetic energy**, provided it is in the form $\frac{mv^2}{2}$. In general, both quantities, momentum and kinetic energy, are conserved, although in special cases one or the other may seem to disappear.

Newton's definition of quantity of motion needs to be qualified in one important way. He should have specified that velocity is a directional quantity. If a body departs from straight line motion, its momentum is thereby changed, even if it remains numerically the same. One reason why this distinction is necessary can be seen from considering a collision between two balls of putty, each having the same mass and speed. In such a collision, the two balls will come to a complete stop. Such a collision would violate the conservation of momentum (mv), i.e., the claim that the quantity of motion defined as momentum, always remains the same. Because, however, momentum is equal to $m\vec{v}$ and hence is itself a directional quantity, the momenta of the two

---

[23]Leibniz's paper in translation has the title "A Short Demonstration of a Remarkable Error of Descartes and Others Who Affirm It to Be a Law of Nature That the Same Quantity of Motion Is Always Conserved by God; Which Law They Make Improper Use of in Applying It to Mechanics." For a translation of it, see William F. Magie (ed.), *Source Book in Physics* (Harvard University Press, 1963), pp. 51–55.

balls will be numerically the same but exactly opposite in direction. Hence, the sum of their momenta, both before and after the collision, will be the same, i.e., zero. Consequently, this collision does not violate the claim that the quantity of motion always remains the same.

Newton's Definitions, continued[24]

Definition 3

The inherent force of matter is the power of resisting, by which each and every body, to the extent that it can, perseveres in its state either of resting or of moving uniformly in a straight line.

This is always proportional to its body, and does not differ in any way from the inertia of mass, except in the mode of conception. Through the inertia of matter it comes to be that every body is with difficulty disturbed from its state either of resting or of moving. Whence the inherent force can also be called by the extremely significant name, "force of Inertia." A body exercises this force only in the alteration of its status by another force being impressed upon it, and this exercise falls under the diverse considerations of resistance and impetus: resistance, to the extent that a body resists an impressed force in order to preserve its state, and impetus, to the extent that the same body, in giving way with difficulty to the force of a resisting obstacle, endeavors to change the state of that obstacle. Common opinion attributes resistance to things at rest and impetus to things in motion, but motion and rest, as they are commonly conceived, are distinguished from each other only with respect [to each other], nor are those things really at rest which are commonly seen as if at rest.

Commentary on Definition 3

Newton's definition of "*Vis Insita*" or "Inherent Force of Matter" or ***Vis Inertiae*** is both quite complex and extremely important. To appreciate what Newton had in mind, recall Aristotle's view that all motion has to be explained. The conceptual framework in which Newton worked was quite different. For him as for Aristotle, rest does not need explanation. By centuries of thought running from Philoponus through Avicenna and Buridan to Galileo and Descartes, scientists had come to see that a body moving at a constant velocity is in a certain sense at rest. The Copernican system had

---

[24] Densmore, *Central Argument* (footnote 19), pp. 9–10.

contributed to the recognition of the relativity of motion, but even more important were such analyses as that given by Galileo of motion on a moving ship.

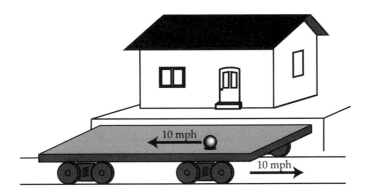

To choose a more contemporary example, imagine a flat-bed train car passing through a station at a rate of 10 mph. Suppose that a person on the train rolls a bowling ball to the rear of the car at a rate of 10 mph. How fast will the ball be moving from the perspective of a person (1) on the train; (2) in the station? The person on the train will see the ball move at 10 mph; to someone at the station, however, the bowling ball will seem to be at rest because the velocities given it by the train and by the bowler exactly cancel each other. This suggests what Newton stresses—that a body moving at a uniform velocity can be viewed as in a state of rest with respect to some reference frame. Consequently, its motion does not need explanation, whereas a change in its motion would.

To deal with this problem, Newton presents the concept **inertia**. But what sort of entity is inertia in the Newtonian system? Is it, for example, a quantity essentially like the impetus of Buridan? Surely not, because bodies have Newtonian inertia even when not moved. On the other hand, Newton categorizes inertia as a force, which is proportional to the mass of the body. Yet in general for Newton, as we shall see, force is what produces change in motion. Moreover, inertia is a strange sort of force; whereas the gravitational force associated with every body continually acts, this force is different; as Newton remarks, "A body exercises this force only in the alteration of its status by another force being impressed upon it...." This analysis suggests that the medieval notion of impetus remained to some extent with Newton. A vast literature exists on this topic; at most, this discussion suggests the type of issues treated in it.

Newton's Definitions, continued[25]

Definition 4

Impressed force is an action exerted upon a body for changing its state either of resting or of moving uniformly in a straight line.

This force consists in the action alone, and does not remain in the body after the action. For the body continues in each new state through the force of inertia alone. Moreover, impressed force has various origins, such as from impact, from pressure, from centripetal force.

Commentary on Definition 4

Newton's notion of **impressed force** is obviously set out in opposition to the force of inertia. As Newton points out, it "does not remain in the body after the action." Examples of such forces would be the force of gravitation or the force produced by a collision.

Newton's Definitions, continued[26]

Definition 5

Centripetal force is that by which bodies are pulled, pushed, or in any way tend, towards some point from all sides, as to a center.

Of this kind is gravity, by which bodies tend to the center of the earth; magnetic force, by which iron seeks a magnet; and that force, whatever it might be, by which the planets are perpetually drawn back from rectilinear motions and are driven to revolve in curved lines. A stone, whirled around in a sling, attempts to depart from the hand that drives it around, and by its attempt stretches out the sling, doing so more strongly as it revolves more swiftly, and as soon as it is released, it flies away. The force contrary to that attempt, by which the sling perpetually draws the stone back to the hand and retains it in its orbit, I call "centripetal," because it is directed towards the hand as to the center of the orbit. And the account of all bodies that are driven in a curved path is the same. They all attempt to recede from the centers of the orbits, and in the absence of some force contrary to that attempt,

---

[25] Densmore, *Central Argument* (footnote 19), p. 12.
[26] Densmore, *Central Argument*, pp. 14–15, 16, 17, 19, 20–21.

by which they are pulled together and kept in their orbits, and which I therefore call "centripetal," they will go off in straight lines with uniform motion.

If a projectile were deprived of the force of gravity, it would not be deflected towards the earth, but would go off in a straight line toward the heavens, doing so with a uniform motion, provided that the resistance of the air be removed. It is drawn back by its gravity from the rectilinear path and is perpetually bent towards the earth, more or less according to its gravity and the velocity of motion. Where its gravity is less in proportion to the quantity of matter, or where the velocity with which it is propelled greater, it will deviate correspondingly less from the rectilinear path, and will travel farther.

If a lead ball, propelled by gunpowder from the summit of some mountain in a horizontal line with a given velocity, were to travel in a curved line for the distance of two miles before it fell to earth, it would travel about twice as far with double the velocity, and about ten times as far with ten times the velocity, provided that the resistance of the air be removed. And by increasing the velocity, the distance to which it is propelled may be increased at will, and the curvature of the line which it describes may be diminished, so that it would finally fall at a distance of ten or thirty or ninety degrees, or it might even go around the whole earth, or, at last, go off towards the heavens, continuing on *in infinitum* with the motion with which it departed. And by the same account, by which a projectile may be deflected into an orbit by the force of gravity and may go around the whole earth, the moon too, whether by the force of gravity (provided it be heavy) or by another force of whatever kind, by which it is urged towards the earth, can be always pulled back towards the earth from its rectilinear path, and deflected into its orbit; and without such a force the moon cannot be held back in its orbit.

This force, if it were less than required, would not sufficiently deflect the moon from the rectilinear path; and if greater than required, would deflect it more than sufficiently, and would lead it down from its orbit towards the earth. It is indeed requisite that it be of exactly the right magnitude, and it is for the Mathematicians to find the force by which a body can be accurately kept back in any given orbit you please with a given velocity, and in turn to find the curvilinear line into which a body departing from any given place you please with a given velocity would be deflected by a given force.

Further, the quantity of this centripetal force is of three kinds: absolute, accelerative, and motive.

Definition 6

The absolute quantity of centripetal force is the measure of the same, greater or less in proportion to the efficacy of the cause propagating it from the center through the encircling regions.

As the magnetic force is greater in one magnet, less in another, in proportion to the size [*moles*] of the magnet or the intensity of the power [*virtus*].

Definition 7

The accelerative quantity of centripetal force is the measure of the same, proportional to the velocity which it generates in a given time.

Thus the power of the same magnet is greater at a less distance, less at a greater one; or the gravitating force is greater in valleys, less on the peaks of high mountains, and less still (as will become clear hereafter) at greater distances from the earth's globe. At equal distances, however, it is everywhere the same, because it equally accelerates all falling bodies (heavy or light, great or small) once the resistance of the air is removed.

Definition 8

The motive quantity of centripetal force is the measure of the same proportional to the motion which it generates in a given time.

Thus the weight is greater in a greater body, less in a lesser one, and in the same body it is greater near the earth, less in the heavens. This quantity is the whole body's centripetency, or propensity towards the center, and (if you will) weight, and is always known through the force contrary and equal to it, by which the descent of the body can be prevented.

[Newton's Commentary to Definitions 6–8]

For the sake of brevity, these quantities of force may respectively be called motive forces, accelerative forces, and absolute forces, and for the sake of distinction, may be said to trace their origins respectively to the bodies seeking the center, to the places of the bodies, and to the center of forces. That is, motive force traces its origin to the body, as if the force were the endeavor of the whole toward the center, composed of the endeavors of all the parts; accelerative force traces its origin to the place of the body, as if it were a sort of efficacy, spread out from the center through the individual places on the circumference, for moving bodies that are in those places; and absolute force traces its origin to the center, as if it were endowed with some cause without which the motive forces would not be propagated through the regions on the circumference, whether that cause be some central body (such as is a magnet at the center of the magnetic force, or the earth in the center of the gravitating force) or something else that does not appear. This concept is strictly mathematical, for I am not now considering the causes and physical seats of the forces.

The accelerative force is accordingly to the motive force as speed is to motion. For the quantity of motion arises from the speed and the quantity of matter, and the motive force arises from the accelerative force and the quantity of the same matter conjoined. For the total action of the accelerative force on the individual particles of a body is the motive force of the whole. Whence, near the surface of the earth, where the accelerative gravity or the gravitating force is the same in all bodies, the motive gravity, or weight, is as the body; but if an ascent be made to regions where the accelerative gravity is less, the weight gradually decreases, and will always be as the body and the accelerative gravity conjoined. Thus in regions where the accelerative gravity is less by a factor of two, the weight of a body that is smaller by a factor of two or three will be less by a factor of four or six.

Further, I call attractions and impulses accelerative and motive in the same sense. Moreover, I use the words "attraction," "impulse," or [words denoting] a propensity of any kind toward a center, indifferently and promiscuously for each other: I am considering these forces, not physically, but only mathematically. Therefore, the reader should beware of thinking that by words of this kind I am anywhere defining a species or manner of action, or a cause or physical account, or that I am truly and physically attributing forces to centers (which are mathematical points) if I should happen to say either that centers attract, or that forces belong to centers.

Commentary on Definitions 5–8

Huygens had previously introduced the notion of **centrifugal** force; by this, he meant the tendency of a body to flee (Latin: *fugere*) from the center. Newton added the term for the opposing force, the **centripetal** force, to describe the force pulling the body toward (Latin: *petere*, "to seek") the center of motion.

It is important to recognize that by the word *motion* in Definition 8 Newton means what we would now call momentum, which, put symbolically, is mv, i.e., the product of the mass and velocity.

Introductory Comment on Newton's Scholium

A scholium is an explanatory note or commentary. Scholia occur at various points in the *Principia*. In this case, Newton treats the vexed topic of **absolute space and time**. Given all the problems with relative motion that we have encountered, we can see that the question whether we can speak meaningfully of an absolute space or time has become all the more difficult. On the one hand, it appears that unless we can locate some absolute frame of reference, it seems problematic whether the Copernican theory can be proven. In addition, by Newton's definition of mass in which weight becomes dependent on location, Newton has in one sense added a relativistic notion to his physics. On the other hand, serious problems seem to stand in the way of designating some frame of reference as absolute, either for space or time. As we shall see, Einstein adopted a position very different from Newton's in regard to the absoluteness of space and time. Nonetheless, Einstein had to erect a new absolute, which suggests that a physics lacking some type of absolute will be conceptually insufficient. As part of his Scholium, Newton proposes an experiment (his bucket experiment) by which he believes it is possible to demonstrate that an absolute frame of reference exists for motion. It is noteworthy that he makes no attempt to establish that there exists an absolute frame of reference for motion of translation (rectilinear motion), but did believe he could do this for rotational motion. Newton's experiment has been widely discussed by physicists and philosophers.

Newton's Scholium to the Definitions[27]

Hitherto it has seemed appropriate to explain the less familiar terms, [and] the sense in which they are to be taken in what follows. Time, space, place, and motion, are very familiar to everyone. It should nevertheless be noted that these are not commonly conceived of otherwise than from their relation to sensible objects. And from this there arise certain prejudices, for the removal of which it is useful for these same [terms] to be distinguished into absolute and relative, true and apparent, mathematical and common.

I. Absolute, true, and mathematical time, in itself and by its nature without relation to anything external, flows uniformly, and by another name is called "duration." Relative, apparent, and common [time] is the perceptible and external measure (whether accurate or varying in

---

[27] Densmore, *Central Argument* (footnote 19), pp. 22–27.

rate) of any duration you please by means of motion, which is commonly used in place of true time, such as an hour, a day, a month, a year.

II. Absolute space, by its nature, without relation to anything external, always remains similar and motionless. Relative [space] is any movable measure or dimension you please of this space, which [measure] is defined by our senses through its position with respect to bodies, and is commonly taken in place of motionless space, such as the dimension of subterraneous, aerial, or celestial space defined through its position with respect to earth. Absolute and relative space are the same in form and size, but do not always remain the same in number. For if the earth, for example, were to move, the space of our air, which relatively and with respect to our earth always stays the same, will be now one part of absolute space in which the air moves across, now another part of it, and thus, absolutely, it will perpetually change.

III. Place is the part of space which a body occupies, and is absolute or relative according to the space. It is a part, I say, of space, not the location of a body, or the enclosing surface. For the places of equal solids are always equal; the surfaces, however, are nearly always unequal because of the dissimilarity of figures. Locations, on the other hand, do not, properly speaking, have quantity, nor are they so much places as properties of places. The motion of the whole is the same as the sum of the motions of the parts: that is, the translation of the whole from its place is the same as the sum of the translations of the parts from their places. Consequently, the place of the whole is the same as the sum of the places of the parts, and for that reason it is internal and in the whole body.

IV. Absolute motion is the translation of a body from absolute place to absolute place; relative [motion is the translation of a body] from relative [place] to relative [place]. Thus in a boat which is carried with sails set, the relative place of a body is that region of the boat in which the body is, or that part of the whole concavity which the body fills, and which to that extent moves along with the boat; and relative rest is the body's continuing to remain in the same region of the boat or part of the concavity. But true rest is the body's continuing to remain in the same motionless part of that space in which the boat itself along with its concavity and all its contents moves. Whence if the earth is really at rest, a body which is relatively at rest in the boat will move truly and absolutely with that velocity with which the boat moves upon the earth. If on the contrary the earth also moves, the true and absolute motion of the body arises partly from the true motion of the earth in motionless space, partly from the relative motion of the boat upon the earth; and if the body also moves relatively in the boat, its true motion arises partly from the true motion of the earth in motionless space, partly from the relative motions of the boat upon earth and of the body in the boat; and from these relative motions there arises the body's relative motion upon earth. Thus, if that part of earth where the boat is placed be really moved eastward with a velocity of 10,010 units, and the boat be carried by the sails and the wind westward with a velocity of ten units, while the boatman walk on the boat towards the east with a velocity of one unit, the boatman will move truly and absolutely in motionless space with 10,001 units of velocity eastward, and relatively upon earth towards the west with nine units of velocity.

Absolute time is distinguished from relative in astronomy by the equation of common time. For the natural days are unequal, but are commonly taken as if equal for a measure of time. Astronomers make a correction for this inequality, so that they may measure the celestial motions from a truer time. It is possible that there is no uniform motion by which time may be measured accurately. All motions can be accelerated and retarded, but the flow of absolute time cannot be changed. The duration or perseverance of the existence of things is the same, whether the motions are fast or slow or none. Furthermore, these are rightly distinguished from their perceptible measures, and are reckoned from them through the astronomical equation. Moreover, the need for this equation in determining phenomena is established both by the experimental evidence of the pendulum clock and by the eclipses of Jupiter's moons.

As the order of the parts of time is unchangeable, so also is the order of the parts of space. If these move from their places, they will also (so to speak) move from themselves. For times and spaces are in a way the places of themselves and of all things. Everything without exception is located in time according to order of succession, in space according to order of place. It is of their essence that they be places, and it is absurd for primary places to move. These are therefore absolute places, and translations from these places are alone absolute motions.

However, because these parts of space cannot be seen and cannot be distinguished from each other by our senses, we introduce perceptible measures in their stead. For from the positions and distances of things from some body, which we see as motionless, we define all places universally, and thereafter we also estimate all motions as well with respect to the places mentioned previously, insofar as we conceive the bodies to be carried away from them. We thus use relative places and motions in place of absolute ones, and this is not an inconvenience in human affairs. In philosophical matters, however, an abstraction from the senses must be made. For it can happen that there is no body really at rest, to which places and motions may be referred.

Further, absolute and relative motions are distinguished from each other through their properties and causes and effects. It is a property of rest that bodies really at rest are at rest among themselves. Therefore, since it is possible that some body in the regions of the fixed stars, or far beyond, be absolutely at rest, while it cannot be known from the position of bodies with respect to each other in our regions whether one of them may preserve its given position with respect to that distant [body], true rest cannot be defined from the position of the latter [bodies] among themselves.

It is a property of motion that the parts, which preserve given positions with respect to the wholes, participate in the motions of the same wholes. For all parts of bodies moving in curves strive to recede from the axis of motion, and the impetus of bodies moving forward arises from the conjoined impetus of the individual parts. Therefore, when the surrounding bodies are moving, those are moving which are at rest among the surrounding ones. And for that reason, true and absolute motion cannot be defined through translation away from nearby bodies, which are viewed as if they were resting bodies. For the external bodies ought not only to be viewed as if they were resting, but also to be truly at rest. Otherwise, all the surrounded [motions], other than translation away from the nearby surrounding [bodies], will also participate in the true motions of the surrounding [bodies], and when that translation is removed, they are not truly at rest but will only be viewed as if they were resting. For the surrounding are to the surrounded as the exterior part of the whole is to the interior part, or as the shell to the kernel. And when the shell is moved, the kernel too is moved without translation away from the surrounding shell, as a part of a whole.

Related to the preceding property is that when the place is moved the thing in the place is moved along with it: thus a body that is moved away from a moved place, also participates in the motion of its place. Therefore, all motions which take place away from [i.e., with respect to] moving places are only parts of the whole and absolute motions, and every whole motion is compounded of the motion of a body from its prime place, and the motion of that place from its place, and so on, until it comes to a motionless place, as in the example of the boatman mentioned above. Hence, whole and absolute motions can be defined only through motionless places, and I have consequently related them above to motionless places, and relative [motions] to movable [places]. However, only those places are motionless which from infinity to infinity all preserve given positions with respect to each other, and moreover remain forever motionless, and constitute that space which I call immovable.

The causes by which true and relative motions are distinguished from each other are the forces impressed upon bodies for generating motions. True motion is neither generated nor changed except by forces impressed upon the moved body itself. But relative motion can be generated and changed without forces being impressed upon this body. For it suffices that they be impressed only upon other bodies to which it is related, so that, when they give way, that relation, of which the relative rest or motion of this [body] consists, is changed. Again, true motion is always changed by forces impressed upon the moved body, but relative motion is not necessarily changed by these forces. For if the same forces were impressed upon other bodies as well, to which there is a relation, in such a way that the relative position be preserved, the relation in which the relative motion consists will be preserved. Therefore, every relative motion can be changed while the true motion is preserved, and can be preserved while the true motion is changed, and for that reason true motion consists not at all in relations of this sort.

The effects by which absolute and relative motions are distinguished from each other are the forces of receding from the axis of circular motion. For in purely relative circular motion, these forces are none, but in true and absolute [circular motion] they are greater or less according to the quantity of motion.

Suppose that a pail should hang from a very long cord, and be driven continually in a circular path, until the cord becomes somewhat stiff from twisting, and [the pail] next be filled with water, and be at rest along with the water, and then be driven by some sudden force in a circular path with a contrary motion, persevering in that motion for a long time as the cord untwists. At the beginning, the surface of the water will be flat, as it was before the motion of the vessel, but after the vessel, by a force gradually impressed upon the water, makes it too begin to rotate perceptibly, it will itself gradually withdraw from the middle, and will climb up to the sides of the vessel, adopting a concave form (as I have myself experienced); and, with an ever increasing motion, will climb more and more, until it make its revolutions in an equal time with the vessel, and come to rest relative to it. This climbing is an indicator of a striving to

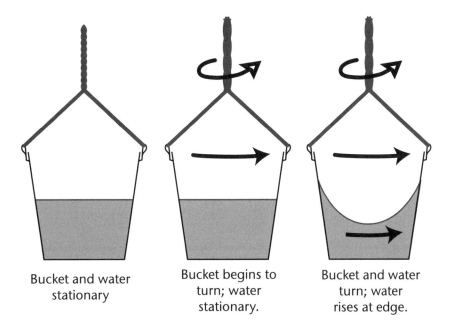

| Bucket and water stationary | Bucket begins to turn; water stationary. | Bucket and water turn; water rises at edge. |

Illustration of Newton's Bucket Experiment

withdraw from the axis of motion, and through such a striving the true and absolute circular motion of the water, here completely contrary to relative motion, comes to be known and is measured. At first, where the relative motion of the water in the vessel was greatest, that motion did not arouse any striving to withdraw from the axis: the water was not seeking the circumference by climbing up the sides of the vessel, but stayed flat, and for that reason its true circular motion had not yet begun. Later, however, where the relative motion of the water decreased, its ascent up the sides of the vessel was an indicator of the striving to withdraw from the axis, and this striving showed its true circular motion, which was ever increasing, and was finally made greatest where the water came to rest relatively in the vessel. Therefore, this striving will not depend upon the translational motion of the water with respect to the surrounding bodies, and consequently, true circular motion cannot be defined by such translations.

The true circular motion of any revolving [body] is unique, corresponding to a unique striving which is its own, as it were, and is commensurate with the effect. Relative motions, however, are countless, in accord with the various relations with external [bodies], and, like relations, they are entirely bereft of true effects, except insofar as they participate in that true and unique motion. Hence also, in the system of those people who would have our heavens rotate in an orb beneath the heavens of the fixed stars, and bear the planets along with them, the individual parts of the heavens, and the planets which are indeed relatively at rest in the heavens nearest them, in truth move. For they change their positions with respect to each other (unlike what happens in those truly at rest), and, being carried along with the heavens, they participate in their motion, and, like the parts of all revolving things, strive to recede from their axes.

Relative quantities are therefore not those quantities themselves whose names they display, but are those perceptible measures of them (whether true or erroneous) which are commonly used in place of the measured quantities. And if the meanings of words should be defined from usage, then by those names "time," "space," "place," and "motion," these perceptible measures should properly be understood, and the discussion will be out of the ordinary and purely mathematical if the quantities measured be understood here. Furthermore, they do violence to sacred scripture who interpret these words as concerning measured quantities there. Nor are mathematics and philosophy any the less defiled by those who confuse the true quantities with their relationships and common measures.

To recognize the true motions of individual bodies and to distinguish them in fact from the apparent ones, is indeed extremely difficult, for the reason that the parts of that motionless space in which the bodies truly move do not flow in to the senses. Nevertheless, the cause is not entirely hopeless. For arguments can be taken, partly from the apparent motions which are the differences of true motions, partly from the forces which are the causes and effects of the true motions.

As, if two globes, joined together at a given distance from each other by a cord between them, were to revolve about their common center of gravity, the striving of the globes to recede from the axis of motion would come to be known from the tension of the cord, and from that the quantity of circular motion might be computed. Then if any equal forces you please were to be simultaneously impressed upon alternate faces of the globes, to increase or decrease the circular motion, the increase or decrease of the motion would come to be known from the increased or decreased tension of the cord, and finally, it would be possible from that to find the faces of the globes upon which forces should be impressed in order most greatly to increase the motion, that is, the aftermost faces, or those which are the following ones in the circular motion. And when the faces which are following are known, and the opposite faces which are leading, the direction of motion would be known. In this way, both the quantity and the direction of this circular motion might be found in whatever immense void you please, where nothing external and perceptible were to exist with which the globes might be compared. Now, if there were to be set up in that space some distant bodies maintaining a given position among themselves, such as are the fixed stars in the regions of the heavens, it would indeed be impossible to tell from the relative translation of the globes within the bodies whether the motion should be attributed to the former or the latter. But if attention were paid to the cord, and it were ascertained that its tension were exactly that which the motion of the globes would require, it would be permissible to conclude that the motion belonged to the globes, and that the bodies were at rest, and, finally, from the translation of the globes within the bodies, to determine the direction of the motion. But how to determine the true motions from their causes, effects, and apparent differences, and, conversely, how to determine their causes and effects from the motions, whether true or apparent, will be taught more fully in what follows. For it is to this end that I wrote the following treatise.

Commentary on Newton's Scholium

Some parts of Newton's discussion of absolute space and time are unproblematic; for example, it is certainly legitimate for him to assert that we should measure time astronomically rather than in terms of some entity such as the length of day. Moreover, it seems clear that Newton avoided making any claim that it is possible to show that translational motion (i.e., motion in a straight line) can be determined in absolute terms. What is striking and very important is that Newton does proceed to claim that rotational motion is not only absolute, but that we can detect its magnitude (his bucket experiment). Overall, Newton asserts that absolute space and time do exist and that we can meaningfully speak of them. In addition, he presents a number of complicated physical arguments in support of this position. Although various important contemporaries of Newton, e.g., G. W. Leibniz, George Berkeley, and Christiaan Huygens disagreed with Newton's claim that it is legitimate to speak of absolute space and time, Newton's views prevailed for over a century and one half. After that, as we shall see subsequently, James Clerk Maxwell, Ernst Mach, Albert Einstein, and others successfully challenged Newton's notion of absolute space and time.

Axioms, or Laws of Motion[28]

Law 1

Every body continues in its state of resting or of moving uniformly in a straight line, except insofar as it is driven by impressed forces to alter its state.

Projectiles continue in their motions except insofar as they are slowed by the resistance of the air, and insofar as they are driven downward by the force of gravity. A top, whose parts, by cohering, perpetually draw themselves back from rectilinear motions, does not stop rotating, except insofar as it is slowed by the air. And the greater bodies of the planets and comets preserve their motions, both progressive and circular, carried out in spaces of less resistance, for a longer time.

Commentary on Law 1: The Law of Inertia

Note: Because Newton's law of inertia is so closely linked conceptually to his definition of inertia, it would be wise to reread at this point his definition of inertia (Def. 3) and the commentary on it.

One question that immediately arises about these axioms is: What, for Newton, is an axiom? In a scholium that follows the presentation of his laws and their corollaries, Newton states that he has in this section "...presented principles accepted by mathematicians and confirmed by manifold experience."[29] In his 1704 *Opticks*, which also begins with a set of axioms, he describes these as "what hath been generally agreed upon."[30] An example of an axiom from his *Opticks* is the law of reflection of light, which had been known since antiquity and which Newton stated as: "The Angle of Reflexion is equal to the Angle of Incidence."[31]

These considerations suggest that Newton may have viewed an axiom as an empirical result that few if any would contest. But this meaning seems problematic if applied to the law of inertia. It is certainly not an empirically obvious result. This is clear from the fact that most authors from Aristotle to the seventeenth century would have denied that it is true. Moreover, if his first law is read as stating that a body will continue *eternally* to move in a straight line unless acted upon by some force, it is hard to see how this

---

[28] Densmore, *Central Argument* (footnote 19), p. 29.

[29] As translated in Densmore, *Central Argument*, p. 39.

[30] Isaac Newton, *Opticks*, 4th ed. (Dover, 1952), p. 20.

[31] Newton, *Opticks*, p. 5.

could be directly proven by observation. Perhaps Newton's axioms should be seen as highly refined hypotheses that can be tested by the conclusions drawn from them. This possibility is supported by the fact that in some of his pre-1687 formulations of mechanics (for example, in his *De motu*), Newton labeled the law of inertia a hypothesis. There are two issues at stake here: one is how Newton viewed the law of inertia and his other axioms; the other is how they should most properly be viewed. Both issues continue to be debated.

Another point to be noted about Newton's first law is that although it tells us how to detect a force—by a change in motion—it does not provide a method of quantifying forces, i.e., we are not told how to determine whether one force is, for example, double another force.

Axioms, or Laws of Motion, continued[32]

Law 2

The change of motion is proportional to the motive force impressed, and takes place following the straight line in which that force is impressed.

If some force should generate any motion you please, a double [force] will generate a double [motion], and a triple [force] a triple [motion], whether it has been impressed all at once, or gradually and successively. And because this motion is always directed in the same way as the generating force, if the body was previously in motion, then this [impressed] motion is either added to its motion (if they have the same sense) or subtracted (if contrary), or joined on obliquely (if oblique) and compounded with it according to the determination of the two.

Commentary on Law 2: The Force Law

Scholars have seen parts of Newton's formulation of his second law as clear and unproblematic; parts, however, have generated controversies among scholars. Let us begin with the former features. Note that for Newton force and motion are directed (in modern parlance, **vectorial**) quantities. To change the direction of a velocity is to change the velocity (even if the speed remains the same); consequently, some force must be proposed as causing the change. This distinction between ordinary quantities and directed quantities had not been made explicit by either Aristotle or Galileo.

---

[32] Densmore, *Central Argument*, p. 29.

On the other hand, over the last sixty years, there has been substantial controversy among scholars in regard to how Newton's formulation of the second law should be understood. This issue concerns what is meant by his phrase "**change of motion.**" Does this mean change of momentum (mv) or does it mean change of momentum with respect to time? Put in the form of an equation, does Newton's second law state $F \propto \Delta mv$ or $F \propto \frac{\Delta mv}{\Delta t}$? Put physically, does Newton view force as proportional to the change (no matter how fast) in the momentum of a body or does he conceive it as proportional to the **rate** at which the momentum changes with respect to time? Put yet another way, is Newton's conception of force in his second law primarily aimed at impulsive forces, e.g., the force acting in the collision of two billiard balls, where the forces act for practical purposes instantaneously? Or is it directed at cases such as a continuously operating gravitational force acting over a period of time? The resolution of this dilemma accepted by most scholars is that in his second law, Newton referred primarily to impulsive forces. He knew, however, how to extend this conception to continuously acting forces, as he did in his first theorem.

In post-Newtonian formulations of mechanics, the second law is usually symbolized as $F = kma$, meaning that the force is proportional to the mass of the body acted on times its acceleration (k being a constant). Note that because acceleration and force are vectorial quantities, they must lie in the same direction. Because $a = \frac{\Delta v}{\Delta t}$, it is clear that this entails the second notion of force, i.e., force as acting continuously. The equation $F = kma$ never appears in this form in the *Principia*, yet Newton in effect used it repeatedly in Book III of his *Principia*.

Although it is clear that Newton intended that his second law be seen as a law, scholars have raised the question of whether, viewed as an equation, the law $F = kma$ can be used to make a definition, in particular, as a way of getting (what Newton was not seeking) a measurement of specific units of force. To see why this question is raised, recall that we have clear definitions of time and distance and, correspondingly, of velocity and acceleration, which are specified in terms of time and distance. But how are mass and force to be measured? One answer is that they are measured in terms of the equation $F = kma$; in other words, force is defined in terms of mass. In this process, scientists designate a particular bar of metal as having a mass of one unit; then other masses are determined in comparison with it. Moreover, forces are determined by the acceleration that such a mass undergoes.

Another question that is worth asking concerning Newton's Law 2 is whether it makes Law 1 unnecessary. The idea is that if no force acts, then no change in motion occurs, or, to put it differently, Law 1 seems to be no more than a statement to cover the special case of $F = kma$, where $F = 0$, which naturally entails that for a body of finite mass, the acceleration a must equal 0.

Problem

Suppose a body of weight w hangs on a scale in an elevator. Suppose the elevator is given a constant velocity v downward. What then will be the weight of the object? Suppose the body is given a constant acceleration a downward such that the scale now reads zero. Determine a.

Solution: Giving the elevator a constant velocity has no effect on the weight of objects in it (Newton's first law). The scale will read w. In the second case, we know $w = kmg$, where m is the body's mass and g its gravitational acceleration. If the elevator accelerates downward in such a way that the scale reads zero, this process must produce an upward force $F = w$. The body's mass does not change. Thus its acceleration must be g, because $kmg = w = F = kma$. If we cancel the m's, we get $a = g = 32 \frac{\text{ft}}{\text{sec}^2}$. From this analysis, it is possible to understand a method that has been used to give astronauts the experience of weightlessness, even without putting them into distant space. The technique consists in putting them in an airplane and letting the airplane for a short period go into free fall. An astronaut in this situation will experience weightlessness, just as he or she would in a region of space where no forces act.

Note on Gravitational versus Inertial Mass

This is a convenient point to discuss another aspect of Newton's notion of mass (Definition One) and simultaneously to introduce an idea that is crucial for understanding Einstein. In defining mass, Newton states that it is proportional to weight. He says that he has proved this by experiments with pendula. What does this mean? There are two quite different ways in which mass manifests itself. It can be manifested as a weight, a force pushing down and caused by gravitational attraction. Call this the body's **gravitational mass**. Also, mass can appear as resistance to motion; think, for example, of the difficulty involved in accelerating a body either in an area of constant gravitational force, e.g., on a sheet of ice, or in a space where no gravitational forces at all are present. In both latter cases, the body resists being accelerated, the measure of this being its **inertial mass**. This raises the question of whether a body's gravitational mass is always equal to the body's inertial mass. If this were the case, then in equations in which both the body's inertial and its gravitational mass appear, it would be unnecessary to designate which is which. Also, it would imply that two bodies composed of different matter but having the same gravitational masses would be equally resistant to acceleration. One way to check this would be to investigate whether a lead and a wooden ball that weigh the same (i.e., have the same gravitational mass) resist acceleration equally. Newton performed just such experiments; his method was to use a hollow pendulum into which he put equal weights (gravitational masses) of different substances; he then checked whether they accelerate identically when swinging. He found that they do.

That nature behaves in this way is very convenient; it is possible to imagine worlds in which a body's gravitational mass does not always equal its inertial mass. In Newton's system, this equivalence is a happy accident. In Einstein's system, it becomes something very different.

Axioms, or Laws of Motion, continued[33]

Law 3

To an action there is always a contrary and equal reaction; or, the mutual actions of two bodies upon each other are always equal and directed to contrary parts.

Whatever pushes or pulls something else is pushed or pulled by it to the same degree. If one pushes a stone with a finger, his finger is also pushed by the stone. If a horse pulls a stone tied to a rope, the horse will also be equally pulled (so to speak) to the stone; for the rope, being stretched in both directions, will by the same attempt to slacken itself urge the horse towards the stone, and the stone towards the horse, and will impede the progress of the one to the same degree that it promotes the progress of the other. If some body, striking upon another body, should change the latter's motion in any way by its own force, the same body (because of the equality of the mutual pushing) will also in turn undergo the same change in its own motion, in the contrary direction, by the force of the other. These actions produce equal changes, not of velocities, but of motions—that is, in bodies that are unhindered in any other way. For changes in velocities made thus in opposite directions, are inversely proportional to the bodies, because the motions are equally changed. This law applies to attractions as well, as will be proven in the next Scholium.

[33]Densmore, *Central Argument* (footnote 19), p. 30.

Commentary on Law 3: Action Equals Reaction

A frequently used statement of Newton's third law is that **for every force, there is an equal and opposite force**. On reflection, this seems obvious enough. As we push down on the floor with our weight, the floor pushes up. As we try to pull in the fish hooked on our fishing line, the fish pulls back. Newton's law consequently seems in some ways as simple as the law that debit and credit must be equal. It is not so simple, however, as the final sentence of Newton's discussion should suggest to you. How can the Earth exert an attractive force on the Moon, which is 243,000 miles away, and why should this force have an equal and opposite force? If the Earth tugs at the Moon with a force F, must the Moon tug the Earth with force F as well? This is what Newton will ask us to believe.

It is an interesting fact that the third law was largely Newton's creation. Simple as it may seem, it has great power when skillfully applied. A host of difficult problems can be solved by means of it, especially when it is used in conjunction with the other laws.

Corollaries to the Laws of Motion[34]

Corollary 1

A body [urged] by forces joined together, describes the diagonal of a parallelogram in the same time in which it describes the sides separately.

Suppose that, in a given time, by the force M alone, impressed at place A, a body would be carried with uniform motion from A to B, and by the force N alone, impressed at the same place, it would be carried from A to C. Let the parallelogram $ABCD$ be completed, and that body

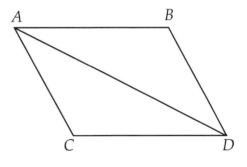

will be carried by both forces in the same time on the diagonal from A to D. For because the force N acts along the line AC parallel to BD, this force, by Law 2, will not at all change the velocity of approach to that line BD generated by the other force. Therefore, the body will arrive at the line BD in the same time whether force N be impressed or

[34] Densmore, *Central Argument* (footnote 19), p. 32.

not; and therefore, at the end of that time, it will be found somewhere on that line BD. By the same argument, at the end of the same time, it will be found somewhere on the line CD, and for that reason it must necessarily be found at the intersection of the two lines D. But, by Law 1, it will proceed with a rectilinear motion from A to D.

Commentary on Newton's Corollaries

In reading Galileo, we encountered the principle of the superposition of velocities; in Corollary 1, we find an analogous and more general idea in Newton. This is the **composition of forces**.

Newton's Corollary 1 is followed by five additional corollaries as well as by a scholium in which Newton urges that a vast number of problems can be solved by means of these laws and their corollaries. The five additional corollaries are quoted below, although without the commentary Newton supplied on them.

Corollary 2: "*And hence is evident the composition of a direct force AD from any oblique forces you please AB and BD, and, in turn, the resolution of any force you please AD into any oblique ones whatever AB and BD. This composition and resolution is, moreover, abundantly confirmed from mechanics.*"

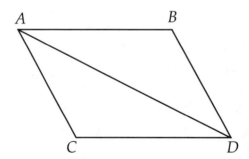

Corollary 3: "*The quantity of motion that is obtained by taking the sum of the motions made in the same direction, and the difference of those made in opposite directions, is not changed by action of the bodies among themselves.*"

Corollary 4: "*The common center of gravity of two or more bodies does not change its state either of motion or of rest by actions of the bodies among themselves, and for that reason the common center of gravity of all bodies acting mutually upon one another (external actions and hindrances being excluded) either is at rest or moves uniformly in a straight line.*"

Corollary 5: *"The motions of bodies contained in a given space are the same among themselves, whether that space be at rest, or whether it move uniformly in a straight line without circular motion."*

Corollary 6: *"If bodies be moved in any manner whatever among them-selves, and be urged by equal accelerative forces along parallel lines, everything goes on moving in the same manner among themselves as if they had not been impelled by those forces."* [35]

Corollary 5 makes clear the degree to which Newton was concerned with the relativity of motion. One commentator in fact describes Corollary 5 as "Newton's Principle of Relativity." [36] Other historians agree; E. J. Dijksterhuis describes it as "the principle of relativity of classical mechanics." [37]

Principia, Book I

On the Motion of Bodies

Section 1[38]

On the method of first and ultimate ratios, by means of which what follows is demonstrated

Lemma 1

Quantities, as well as the ratios of quantities, which in any finite time you please constantly tend towards equality, and before the end of that time approach nearer to each other than by any given difference you please, become ultimately equal.

If you deny it, let them become ultimately unequal, and let their ultimate difference be D. Therefore, they cannot approach nearer to each other than by the given difference D, contrary to the hypothesis.

---

[35] Dana Densmore, *Newton's* Principia: *The Central Argument*, with translations by William H. Donahue, pp. 32–34, 38–39.

[36] Banesh Hoffmann, "Relativity" in Philip Weiner (ed.), *Dictionary of the History of Ideas*, vol. 4 (Charles Scribner's Sons, 1973), p. 76.

[37] E. J. Dijksterhuis, *The Mechanization of the World Picture* (Oxford University Press, 1961), p. 468.

[38] Densmore, *Central Argument* (footnote 19), pp. 47, 50–51, 55.

Lemma 2

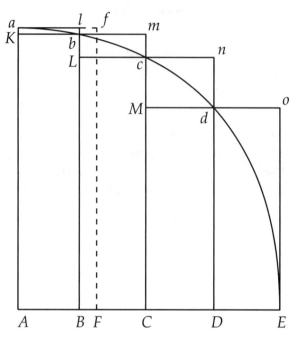

If in any figure you please AacE, contained by the straight lines Aa, AE, and the curve acE, there be inscribed any number of parallelograms Ab, Bc, Cd, and so on, contained beneath equal bases AB, BC, CD, and so on, and by the sides Bb, Cc, Dd, and so on, parallel to the side Aa of the figure; and if the parallelograms aKbl, bLcm, cMdn, and so on, be completed, and if afterward the breadth of the latter parallelograms be diminished and their number be increased in infinitum, *I say that the ultimate ratios which the inscribed figure AKbLcMdD, the circumscribed figure AalbmcndoE, and the curvilinear figure AabcdE, have to each other are the ratios of equality.*

For the difference of the inscribed figure and the circumscribed [one] is the sum of the parallelograms *Kl, Lm, Mn, Do,* that is (because of the equal bases of them all), the rectangle beneath the base *Kb* of one of them and the sum of the altitudes *Aa,* that is, the rectangle *ABla.* But because the breadth *AB* of this rectangle is diminished *in infinitum,* the rectangle becomes less than any given [magnitude] you please. Therefore (by Lemma 1) the inscribed figure and the circumscribed [one], and much more so the intermediate curvilinear [figure], become ultimately equal.   Q.E.D.[39]

Lemma 3

The same ultimate ratios are also ratios of equality, where the breadths of the parallelograms AB, BC, CD, and so on, are unequal, and all are diminished in infinitum.

---

[39]Q.E.D. is an abbreviation for *quod erat demonstrandum,* "that which was to have been demonstrated." [translator's note]

For let AF be equal to the greatest breadth, and let the parallelogram $FAaf$ be completed. This will be greater than the difference of the inscribed figure and the circumscribed figure; and when its breadth AF is diminished *in infinitum*, it becomes less than any given rectangle.   Q.E.D.

Commentary on Newton's Lemmas

Note: Newton's Lemmas 4–11 are not included in this selection from Newton's *Principia*.

The basic idea involved in Newton's lemmas can be illustrated by a simple example. Let us use the method described in Lemma 2 to determine the area of a right triangle, the perpendicular sides of which are a and b. In the proof, we shall need the well known formula for the sum of the integers from 1 to n. This formula is that the sum is equal to $\frac{n(n+1)}{2}$. For example, the sum of the integers from 1 to 6 is $\frac{6(7)}{2} = \frac{42}{2} = 21$.

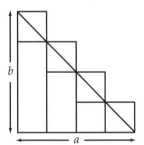
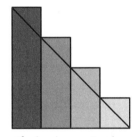
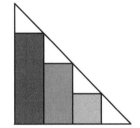

The four large rectangles   The four small rectangles

First, let us attempt to estimate the area A of the triangle. To do this, divide the side a into 4 equal parts of length and erect on these four lines four rectangles that "circumscribe" the triangle (second diagram) and four rectangles that are "inscribed" within the triangle (third diagram). Note that the last rectangle on the right is of zero area.

By inspection of these figures, it is evident that A is less than the sum of the area of the four larger rectangles, but greater than the area of the four smaller rectangles. We can put this in the form of the following equation in which the four quantities on the left side represent the areas of the four rectangles in the middle diagram, and the four quantities on the right side of the equation represent the areas of the four rectangles on the right side.

$$\frac{a}{4} \cdot \frac{4b}{4} + \frac{a}{4} \cdot \frac{3b}{4} + \frac{a}{4} \cdot \frac{2b}{4} + \frac{a}{4} \cdot \frac{1b}{4} > A > \frac{a}{4} \cdot \frac{3b}{4} + \frac{a}{4} \cdot \frac{2b}{4} + \frac{a}{4} \cdot \frac{1b}{4} + \frac{a}{4} \cdot \frac{0b}{4}.$$

Simplifying the left side of the equation, using the previously noted formula for summing a set of consecutive integers from 1 to n, makes the left side equal:

$$\frac{ab}{16}(4+3+2+1) = \frac{ab}{16}\frac{4(5)}{2} = \frac{5ab}{8}.$$

The sum for the right side is:

$$\frac{ab}{16}(3+2+1+0) = \frac{ab}{16}\frac{3(4)}{2} = \frac{3ab}{8}.$$

This tells us that the area of the triangle will lie between $\frac{5ab}{8}$ and $\frac{3ab}{8}$. Let us now apply this same procedure of constructing rectangles, but this time constructing n rectangles. This will result in the following equation:

$$\frac{a}{n}\left\{\frac{nb}{n} + \frac{(n-1)b}{n} + \cdots + \frac{[n-(n-1)]b}{n}\right\} > A >$$
$$\frac{a}{n}\left\{\frac{(n-1)b}{n} + \cdots + \frac{[n-(n-1)]b}{n} + \frac{(n-n)b}{n}\right\}.$$

Simplifying this by exactly the same method used previously produces the equation:

$$\frac{ab}{n^2}\left[\frac{n(n+1)}{2}\right] = \frac{ab}{n^2}\left(\frac{n^2+n}{2}\right) = \frac{ab}{2}\left(1+\frac{1}{n}\right) > A >$$
$$\frac{ab}{n^2}\left[\frac{(n-1)n}{2}\right] = \frac{ab}{2}\left(\frac{n^2}{n^2} - \frac{n}{n^2}\right) = \frac{ab}{2}\left(1-\frac{1}{n}\right).$$

In short,

$$\frac{ab}{2}\left(1+\frac{1}{n}\right) > A > \frac{ab}{2}\left(1-\frac{1}{n}\right).$$

We could plug very large values of n into this equation so as to determine ever more exactly the range of values of A. We can attain a far more interesting result by using Newton's first and second lemmas. Let n go to infinity. As this occurs, both the left and right sides of the equation can be made as close to $\frac{ab}{2}$ as one wishes; in particular, their difference from A can be made smaller than any specified quantity. Consequently, $\frac{ab}{2}$ must be the area of the triangle. This is, admittedly, not a surprising result. What is significant is that this same method can be used to determine the exact areas of more complicated figures, for example, the area of an ellipse or the area under a portion of a parabola.

It is well worth noting that these three lemmas develop ideas that were central to the calculus. As such they illustrate the point that although Newton did not directly introduce the calculus in his *Principia*, some key ideas of the calculus do make their appearance in his book.

Section 2[40]

On the Finding of Centripetal Forces

Proposition 1

The areas which bodies driven in orbits [gyros] describe by radii drawn to an immobile center of forces, are contained in immobile planes and are proportional to the times.

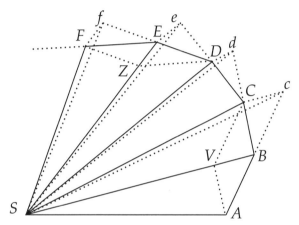

Let the time be divided into equal parts, and in the first part of the time let the body, by its inherent force, describe the straight line AB. In the second part of the time, the same body, if nothing were to impede it, would pass on by means of a straight line to c (by Law 1), describing the line Bc equal to AB, with the result that, radii AS, BS, cS being drawn to the center, the areas ASB, BSc would come out equal. But when the body comes to B, let the centripetal force act with an impulse that is single but great, and let it have the effect of making the body depart from the straight line Bc and continue in the straight line BC. Let cC be drawn parallel to BS, meeting BC at C; and, the second part of the time being completed, the body (by Corollary 1 of the Laws) will be located at C, in the same plane as the triangle ASB.

Connect SC, and, because of the parallels SB Cc, triangle SBC will be equal to triangle SBc, and therefore also to triangle SAB. By a similar argument if the centripetal force should act successively at C, D, E, and so on, making the body describe the individual straight lines CD, DE, EF, and so on, in the individual particles of time, all these will lie in the same plane, and the triangle SCD will be equal to the triangle SBC, and SDE to SCD, and SEF to SDE. Therefore, in equal times

---
[40] Densmore, *Central Argument* (footnote 19), p. 123.

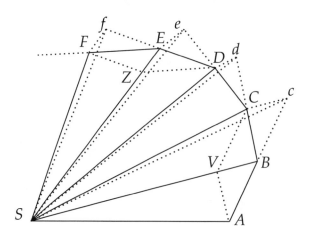

equal areas are described in a motionless plane; and, *componendo*, any sums whatever of areas *SADS, SAFS* are to one another as are the times of description. Now let the number of the triangles be increased and their breadth decreased in infinitum, and their ultimate perimeter *ADF* (by Corollary Four of Lemma Three) will be a curved line: and therefore the centripetal force, by which the body is perpetually drawn back from the tangent of this curve, will act without ceasing, while any described areas whatever *SADS, SAFS*, always proportional to the times of description, will be proportional to those same times in this case.

Q.E.D.

Commentary on Proposition 1

Let us first look at the physics of this demonstration. Newton has shown that if some body is drawn by an immobile center of force, then the body will move in such a way that it sweeps out equal areas in equal times. It is very important to understand that Newton does not specify how that force varies with distance (e.g., inversely as the distance squared, inversely as the distance) nor does he make any mention of the nature of that force (gravitational, electrical, magnetic, etc.). Nor does he state that there must be some body at the center or source of that force; he simply describes it as a centripetal force, i.e., a force directed toward a center. His contemporaries must have been very surprised that Newton would base a proposition on such a minimal assumption, and even more surprised that from the proposition Newton erected on it, he succeeded in demonstrating that Kepler's second law must be the case! No one at Newton's time seems to have suspected that Kepler's second law follows from such a minimal assumption.

Concerning the mathematics of the demonstration, three points deserve comment. First, Newton needs to show that if from point S, lines be drawn to a line ABc on which $AB = Bc$, then area of $\triangle ABS$ = area of $\triangle BcS$. But this is a theorem of Euclid: *"Triangles which are on equal bases and in the same parallels are equal to one another [in area]"* (Bk. I, Prop. 38). Second,

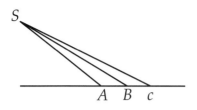

Newton needs to have evidence that $\triangle SBC$ in his diagram is equal in area to $\triangle SBc$. Because Cc is parallel SB, this (again by Euclid, I, 38) must also be the case. Third, Newton must support his claim that if the triangles shrink "in infinitum" in size, the resulting figure will be a curve. This is covered, as he notes, by Corollary 4 of his Lemma 3 of Section 1.

Corollaries to Proposition I.1[41]

I.1 Corollary 1

The velocity of a body attracted to an immobile center in nonresisting spaces is inversely as the perpendicular dropped from that center to the rectilinear tangent of the orbit. For in those places A, B, C, D, E, the velocity is as the bases of the equal triangles AB, BC, CD, DE, EF, and these bases are inversely as the perpendiculars dropped to them.

I.1 Corollary 2

If the chords AB, BC of two arcs described successively in equal times by the same body in nonresisting spaces be completed into the parallelogram $ABCV$, and its diagonal, in that position which it ultimately has when those arcs are diminished in infinitum, be produced in both directions, it will pass through the center of forces.

I.1 Corollary 3

If the chords AB, BC, and DE, EF of arcs described in equal times in nonresisting spaces be completed into parallelograms $ABCV$, $DEFZ$, the forces at B and E are to each other in the ultimate ratio of the diagonals BV, EZ, where these arcs are diminished in infinitum. For the motions BC and EF of the body are (by Corollary 1 of the Laws) composed of the motions Bc, BV, and Ef, Ez; and BV and Ez, equal to Cd and Ff, were in the demonstration of this proposition generated by the impulses of the centripetal force at B and E, and therefore are proportional to these impulses.

---

[41] Densmore, *Central Argument* (footnote 19), pp. 131, 134, 136.

Commentary on the Remaining Sections of Book I

Section 2 of the Book I of the *Principia* consists of ten propositions, some of which are among the most important in the *Principia*. It begins, as we have seen, with a dramatic demonstration of a theorem that Newton himself discovered. The theorem reveals a conclusion about how a body moves when acted on by a force directed to some point, irrespectively of what force law (e.g., an inverse square law) governs that force.

Propositions 2–4

Newton's second proposition is a converse of his first proposition. It shows that if a radius be drawn from body A to body B, and if as B moves, this radius sweeps out equal areas in equal times, then the body B is drawn to that point by a **centripetal** force. We can pass over Proposition 3, but Proposition 4 deserves quoting and commentary. It states:

> *The centripetal forces of bodies that describe different circles with uniform motion tend towards the centers of the same circles, and are to one another as the squares of arcs described in the same times applied to [i.e., divided by] the radii of the circles.*[42]

This is the law of centripetal force, from which by means of Newton's second law one can easily derive the law of centripetal acceleration. If we take v as the velocity of the body, r as the radius of its circular path, and F as the centripetal force, then we may state Proposition 4 as $F \propto \frac{v^2}{r}$. Combining this with Newton's second law of motion, we have the law of centripetal acceleration: $a \propto \frac{v^2}{r}$. Newton had derived this formula in the latter half of the 1660s.

The proof that Newton provided for Proposition 4 is quite complex. Nonetheless, one can get a good sense of Newton's methodology in his proof by following a simpler proof formulated by William H. Donahue.

Quasi-Newtonian Proof of the Law of Centripetal Acceleration[43]

Imagine a body moving at a constant speed on a circle with center S and radius r. Let it begin to move from point A. If no force were acting on the body, it would, by the law of inertia, move in a straight line tangent to the circle. Call that line AE, and after a very short time interval Δt let the body, on the no-force path, arrive at the point E.

---

[42] Densmore, *Central Argument* (footnote 19), p. 154.

[43] I deeply appreciate Dr. William H. Donahue's willingness to allow me to use this proof he created of the law of centripetal acceleration.

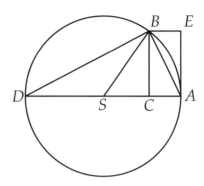

Since the body is really moving on the circle, however, it is being constantly deflected from the tangent path and pulled toward the center. Suppose that after time Δt it arrives at point B. Because we have made Δt very short, the line from E to B may be taken as parallel to AS (the arc AB is drawn rather large here for the sake of clarity; it is to be understood as being very short). Also, AE (being a tangent) is perpendicular to SA, so that $AEBC$ is a rectangle, and line AC is equal to line EB.

Now line EB is the difference between where the body would be if no force acted, and where the body is when acted on by the force toward S, during the time interval Δt. Therefore, it is the distance through which the force moves the body in time Δt. Because the time interval is very short, this force can be taken as constant (in fact, for circular motion, it does remain constant). Therefore, by Galileo's equation for uniformly accelerated motion,

$$EB \text{ or } AC = \frac{a\Delta t^2}{2}, \tag{1}$$

where a is the acceleration produced by the force.

Now let's look at the triangles ABC and ABD. Euclid proved (Bk. III, Prop. 31) that any triangle that is inscribed in a semicircle (such as ABD here) has a right angle on the circumference (that is, at B). And since ABC is also a right triangle, and since the acute angles of any right triangle add up to 90 degrees (Euclid I.32), these two triangles have the same three angles, and are all the same shape: they are similar triangles.

When geometrical figures are similar, their sides are proportional, so that if (for example) AB were twice, three times, etc., as long as AC, then AD (in triangle ABD) would also be twice, three times, etc., as long as AB. We can express this in fractions:

$$\frac{AC}{AB} = \frac{AB}{AD}.$$

If we multiply both sides of the equation by AB, we have

$$AC = \frac{AB^2}{AD}.$$

But because the time is very short, line AB is only imperceptibly different from the arc AB, and $AD = 2r$. Therefore,

$$AC = \frac{\text{arc } AB^2}{2r}.$$

Now the body is moving with constant speed, and so by the familiar distance/speed relation,

$$\text{arc } AB = v\Delta t,$$

where v is the speed. Substituting this into the previous equation,

$$AC = \frac{v^2 \Delta t^2}{2r}.$$

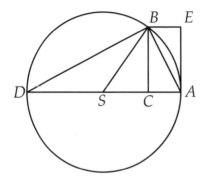

Combining this result with Equation 1, we get

$$\frac{a\Delta t^2}{2} = \frac{v^2 \Delta t^2}{2r}.$$

Removing the common factor $\frac{\Delta t^2}{2}$ from both sides, we have:

$$a = \frac{v^2}{r}.$$

Newton's Corollaries to Proposition 4

Newton adds nine corollaries and a short scholium after Proposition 4. For present purposes, mention need be made only of Corollary 6, which is mathematically essentially identical to the proof (discussed earlier) that Newton formulated in the late 1660s showing (on the assumption that planets move in circular orbits) that from the law of centripetal acceleration and Kepler's third law, one can infer an inverse square field of force. To put this point somewhat differently, Newton had come to see a linkage between Kepler's third law and the inverse square law for a simplified situation (circular motion). This should not be taken to mean that Newton was claiming at this point to have proved the law of universal gravitation.

Propositions 5–15

The propositions between 4 and 11 are mainly directed at laying the groundwork for Proposition 11, which states that if a body moves in an ellipse with the center of force at one focus of the ellipse, then the moving body will accelerate towards the center of force with an acceleration inversely proportional of its distance from that center. This may be rephrased to state that if Kepler's first law (planets move in ellipses with the Sun at one focus) is true, then the inverse square law governs gravitational attraction. Newton does not at this point argue for Kepler's first law; this he will do in Book III. It is interesting and important to note that Newton at this stage refrains from considering the effect of a body at the center of force, leaving this complication for later discussion. Unfortunately, the proof of Newton's Proposition 11 is extremely complex; we shall not attempt it.

Proposition 11 is the first of the seven propositions in Section III. These combine to prove another new and striking result. Because it is possible to derive the inverse square law from Kepler's first law, one might assume that from the inverse square law, elliptical orbits must follow. Newton shows that this is not correct; what is true is that if a body moves in any conic section (ellipse, hyperbola, or parabola) with a center of force at one focus of the conic section, then the inverse square law governs that motion. Put in concrete terms, this suggests that there can be bodies moving around the Sun not only in ellipses, but also in parabolas or hyperbolas. Newton will later argue that comets move in parabolas or hyperbolas. Another side result from this is support for Galileo's discovery that projectiles on Earth move in parabolas.

Proposition 15 reveals another surprising result. It states that "the periodic times in ellipses are in the sesquiplicate ratio of the major axes."[44] This means that if the inverse square force law is true, then for a body moving in an ellipse around another body located at one focus of the ellipse, the period of the body will be proportional to the $3/2$ power of the major axis of the ellipse, i.e., of the length of the longest diameter of the ellipse. It is important to understand the significance of Proposition 15. Recall that we saw earlier in this chapter that Newton around 1669 had made a major step toward the inverse square law when he carried out a calculation that involved combining Kepler's third law with the law of centripetal acceleration (as well as the more or less justifiable assumption that planets move in circular orbits). When this was done, Newton was able to conclude that the acceleration of a planet obeying Kepler's third law toward the center of its circular orbit must be inversely proportional to the square of the planet's distance from the center of its orbit. This suggests that very early on in his thinking about planetary motion, he suspected that the force involved must vary inversely as the square of the distance from the center. In Proposition 15, Newton proceeds in a different direction: he is able to show that from the geometry of the ellipse and the inverse square law, one could derive Kepler's third law.

Newton's presentation up to this point of his *Principia* must have deeply impressed mathematically sophisticated readers. It was becoming apparent

---

[44] Densmore, *Central Argument* (footnote 19), p. 263.

that Newton had managed to express all three of Kepler's laws in terms of his physics of central forces, and that he had shown in particular how the elliptical form of the orbit (Kepler's first law) is closely related to the Keplerian period/distance proportion (Kepler's third law). To put it differently, it was as if the pieces of a giant jigsaw puzzle were beginning to fit together in a remarkable manner. Even this does not quite accurately describe the nature and magnitude of Newton's achievement. Recall that Newton had shown that a body under an inverse square force might move not only in an ellipse, but also in a parabola or in hyperbola (the three types of conic sections). Thus the perspective he offered was even more stunning than, say, an explanation of why planets move in ellipses; he could also explain by the theoretical structure he had created the possibility of other orbital paths. Similarly, Newton had worked out the consequences for the motions of bodies not only in an inverse square field but also for motions in other fields of force, should such exist. In one sense, Newton had worked out a system that could apply to an array of possible universes, which of course left the problem of deciding which of those universes we inhabit. This was one of the goals of Newton's Book III.

The Remainder of Book I of the *Principia*

Sections 1–3 are the core of Book I; secondary results, derived from these early materials, make up most of the rest of its remaining eleven sections. One exception to this generalization is Proposition 75 and the theorems leading up to it. Proposition 75 is the famous theorem in which Newton shows that for two spherical bodies attracting each other, the forces between them act just as if the mass of each body were concentrated at its center. To be more precise, Newton shows that assuming the force of attraction between any two particles varies inversely as the distance between them squared, then for two spherical bodies each composed of particles that attract the particles of the other sphere, the force of attraction between the two spheres as a whole is inversely as the distance between the centers of the two spheres. Newton's statement of the proposition is:

> If to the individual points of a given sphere there tend equal centripetal forces decreasing in the duplicate ratio of the distances from the points, I say that any other similar sphere you please is attracted by the same [sphere] with a force inversely proportional to the square of the distance of the centers.[45]

---

[45] Densmore, *Central Argument* (footnote 19), p. 299.

It is important to keep in mind that Book I of the *Principia* should be seen as primarily a treatise on mathematical aspects of mechanics rather than on the solar system. After completing it, Newton still faced the task of showing that the propositions developed in it can be applied to actual celestial and terrestrial motions. To this task Newton set himself in writing Book III, "The System of the World." To it, we now turn.

Newton's *Principia*, Book III[46]

On the System of the World. Third Book.

[Newton's Preface to Book III]

In the preceding books I have presented the principles of philosophy, not, however, philosophical, but only mathematical; that is, those from which one can argue in philosophical matters. These are the laws and conditions of motions and forces, which pertain to philosophy in the highest degree. Nevertheless, so that these should not appear sterile, I have illustrated them with certain philosophical scholia, treating those things that are most general, and in which philosophy seems most to be poured out, such as the density and resistance of bodies, spaces void of bodies, and the motion of and of sounds. What remains is that we show from the same principles the constitution of the system of the world. Concerning this theme I had composed the third book using a popular method, so that it might be read by many. But those by whom the established principles might not have been sufficiently well understood will have a very slight perception of the force of the argument, and will not cast aside prejudices to which they have been accustomed for many years. Therefore, so that the matter not be subject to dispute, I have carried over the substance of that book into propositions, in the mathematical manner, so that they may only be read by those who had previously gone through the principles. Nevertheless, because many propositions appear there [i.e., in the first two books] which might take too much time even for those learned in mathematics, I do not wish to propose that everyone read all of them: it would be sufficient for one to read carefully the definitions, laws of motion, and the first three sections of the first book, and then to carry on to this book on the system of the world, looking up at will the remaining propositions of the previous books cited here.

---

[46] Densmore, *Central Argument* (footnote 19), pp. 301, 303–305.

Rules of Philosophizing

Rule 1

That there ought not to be admitted any more causes of natural things than those which are both true and sufficient to explain their phenomena.

Philosophers state categorically: Nature does nothing in vain, and vain is that which is accomplished with more that can be done with less. For Nature is simple, and does not indulge herself in superfluous causes.

Rule 2

Accordingly, to natural effects of the same kind the same causes should be assigned, as far as possible.

As, for example, respiration in humans and in animals, the descent of stones in Europe and in America, light in a cooking fire and in the sun, the reflection of light on earth and in the planets.

Rule 3

The qualities of bodies that do not suffer intensification and remission, and that pertain to all bodies upon which experiments can be carried out, are to be taken as qualities of bodies universally.

For the qualities of bodies are apprehended only through experience, and are accordingly to be declared general whenever they generally square with experiments; and those which cannot be diminished cannot be removed. It is certain that against the tenor of experiments, dreams are not to be rashly contrived, nor is a retreat to be made from the analogy of nature, since she is wont to be simple and ever consonant with herself. The extension of bodies is apprehended only through the senses, nor is it perceived in all things. But because it belongs to all perceptible bodies, it is affirmed to be universal. We experience many bodies to be hard. Hardness of the whole, moreover, arises from hardness of the parts, and thence we rightly conclude that the undivided particles, not only of these bodies which are perceived, but also of all others, are hard. We conclude that all bodies are impenetrable, not by reason, but by perception. Those that we handle are found to be impenetrable, and thence we conclude that impenetrability is a property of bodies universally. That all bodies are movable, and that by certain forces (which we call the forces of inertia) they persevere in motion or rest, we gather from these very properties in bodies that we see. Extension, hardness, impenetrability, movability, and the force of inertia of the whole arise from the extension, hardness, impenetrability, movability, and forces of inertia of the parts, and thence we conclude that all the least parts of all bodies are extended and hard and impenetrable and movable and endowed with forces of inertia. And this is the foundation of all of philosophy. Moreover, that parts of bodies that are divided and mutually contiguous can be separated from each other, we come to know from the phenomena, and that undivided parts can by reason be divided up into smaller parts, is certain from mathematics. But whether those distinct and hitherto undivided parts can be divided by the forces of nature and separated from each other, is uncertain. And if it were to be established by but a single experiment that by breaking a hard, solid body, some undivided particle were to suffer division, we would conclude, by the force of this rule, not only that the divided parts are separable, but also that the undivided ones can be divided *in infinitum*.

Finally, if it be established universally by experiments and astronomical observations that all bodies on the surface of the earth are heavy towards the earth, and this according to the quantity of matter in each, and that the moon is heavy towards the earth according to the quantity of its matter, and that our sea in turn is heavy towards the moon, and that all the planets are heavy towards each other, and that the gravity of comets towards the sun is similar, it will have to be said, by this rule, that all bodies gravitate towards each other. For the argument from phenomena for universal gravitation will be even stronger than that for the impenetrability of bodies, concerning which we have absolutely no experiment in the heavenly bodies; nay, not even an observation. Nevertheless, I do not at all assert that gravity is essential to bodies. By "inherent force" [*vis insita*] I understand only the force of inertia. This is inalterable. Gravity is diminished in receding from earth.

Rule 4

In experimental philosophy, propositions gathered from the phenomena by induction are to be taken as true, whether exactly or approximately, contrary hypotheses notwithstanding, until other phenomena appear through which they are either rendered more accurate or liable to exceptions.

This must be done lest an argument from induction be nullified by hypotheses.

Commentary on Newton's "Rules of Philosophizing"

That by "philosophizing" in the title of this section Newton meant natural philosophy or what we call science should be clear from a rapid reading of his rules; that these rules embody an empiricist methodology should also be evident. To appreciate, however, precisely what Newton was attempting to accomplish by means of his formulation of these rules requires some knowledge of the context in which he wrote them. This will show that they were composed not as an interesting digression in an otherwise technical volume, but rather that they play an integral role in his overall argument.

Seventeenth-century scientists were wary of explanations in terms of what were called **occult** qualities. One must not, they insisted, explain the action of, for example, sleeping potions by ascribing to such drugs a dormative or soporific power. This, they urged, would be to try to explain something simply by giving a name to the supposed cause. Aristotelians, it was claimed, were especially guilty of this error. The tendency to invoke such modes of explanation has by no means disappeared. Sports commentators regularly explain why a basketball player scored thirty points by saying that he/she had a hot hand.

Not only was Newton opposed to occult explanations; he was also critical of the *a priori* character of Cartesian science. He believed that Descartes lacked reliable empirical evidence for his claim that vortices of invisible matter fill space and push the planets through their orbits. Consequently, Newton's four rules can be seen as directed in large measure against the Cartesians whom he accused of presenting "dreams."

The situation was, however, more complex than this: Newton was not only *attacking* the dominant continental Cartesian approach, he was also *defending* his own system. Newton's fundamental mode of explanation was, after all, open to attack on at least two grounds. First, gravitational attractions cannot be observed in any direct sense. We understand why the sofa moves when we push it, but how can the Earth exert a force on the Moon, which is 243,000 miles from it? Second, gravity itself was described by some as an occult cause. Such continental quasi-Cartesians as Christiaan Huygens objected that gravity is no explanation, but only the name for an effect. Gravity itself, Huygens insisted, requires explanation. We see then that Newton's rules should be understood as in one sense an attempt to justify his own physical system against such criticisms. In light of this analysis, you may wish to reread Rule 3.

The anti-Cartesian character of Newton's rules has long been recognized. This aspect was pointed out by, for example, **William Whewell** (1794–1866), who in his *Philosophy of the Inductive Sciences* presented a helpful analysis of them. Whewell prefaced his analysis by stating:

In considering these Rules, we cannot help remarking, in the first place, that they are constructed with an intentional adaptation to the case with which Newton has to deal,—the induction of Universal Gravitation; and are intended to protect the reasonings before which they stand. Thus the first Rule is designed to strengthen the inference of gravitation from the celestial phenomena, by describing it as a *vera causa*, a true cause; the second countenances the doctrine that the planetary motions are governed by mechanical forces, as terrestrial motions are; the third rule appears intended to justify the assertion of gravitation, as a *universal* quality of bodies; and the fourth contains, along with a general declaration of the authority of induction, the author's protest against hypotheses, levelled at the Cartesian hypotheses especially.[47]

Newton faced yet another problem. The ability of *both* the Ptolemaic and Copernican systems to save the phenomena suggested that any given set of phenomena could be explained in a number of different ways. Newton was, in fact, aware that many of the phenomena he will be explaining in Book III by gravitational theory could possibly be accounted for very differently by the Cartesians of the continent. In fact, in his paper titled "An Hypothesis Explaining the Properties of Light," which he submitted to the Royal Society in 1675 (first published in the eighteenth century), and in his unpublished 1679 *Letter to Boyle*, Newton himself presented aetherial theories of gravity. To overcome this source of problems, he formulated his Rule 4 and in general stressed the inductive aspects of his system. Another indication of the problem faced in proving various parts of his system comes from a consideration of the law of inertia. The history of this law suggests that the law of inertia is not a result that can be directly and easily induced from observation of phenomena. Moreover, it is a claim about how a body will move *eternally* in a straight line if *no* forces act on it. To remove all forces from acting on a body and then to observe it moving eternally are conditions that no experimentalist would welcome. In what follows, it will be important to watch how Newton attempts to get around such problems as these.

This analysis of the methodological problems that beset Newton helps set the stage for an understanding of the scientific approach used in Book III. Newton begins by setting out various phenomena of nature; he then seeks to show that the mathematical laws and principles developed in the previous books explain these phenomena. With this in mind, let us turn to the scientific matters with which Book III opens.

---

[47] William Whewell, *The Philosophy of the Inductive Sciences, Founded upon Their History*, new ed., vol. 2 (John W. Parker, 1847), p. 279.

Phenomena[48]

Phenomenon 1

That the planets around Jupiter,[49] by radii drawn to the center of Jupiter, describe areas proportional to the times, and that their periodic times (the fixed stars being at rest) are in the sesquiplicate ratio of the distances from Jupiter's center.

This is established from astronomical observations. The orbits of these planets do not differ perceptibly from circles concentric about Jupiter, and their motions in these circles are found to be uniform. But that the periodic times are in the sesquiplicate ratio of the semidiameters of the orbits, the astronomers are in agreement, and this same thing is manifest from the following table.

Periodic Times of Jupiter's Satellites

1d 18h 27m 34s	3d 13h 13m 42s	7d 3h 42m 36s	16d 16h 32m 9s

Distance of the Satellites from the Center of Jupiter

From the Observations	1	2	3	4	
of Borelli	5	8	14	24	
of Townley, by micrometer	5.52	8.78	13.47	24.72	Semidiam.
of Cassini, by telescope	5	8	13	23	of Jupiter
of Cassini, by eclipse of satell.	5	9	14	25	
From the Periodic Times	5.667	9.017	14.384	25.299	

Mr. Pound has determined the elongations of Jupiter's satellites and its diameter by means of the best micrometers as follows. The greatest heliocentric elongation of the fourth satellite from the center of Jupiter was taken by a micrometer in a tube fifteen feet long, and came out to be about 8′16″ at Jupiter's mean distance from earth. That of the third satellite was taken by a micrometer in a telescope 123 feet long, and came out to be 4′42″ at the same distance of Jupiter from earth. The greatest elongations of the remaining satellites at the same distance of Jupiter from earth came out to be 2′56″47‴ and 1′51″6‴, from the periodic times.

The diameter of Jupiter was taken more frequently by a micrometer in a telescope 123 feet long, and, reduced to Jupiter's mean distance from the sun or earth, always came out less than 40″, never less than 38″, and most often 39″. In shorter telescopes this diameter is 40″ or

---

[48] Densmore, *Central Argument* (footnote 19), pp. 308–309.
[49] The satellites or moons of Jupiter. [MJC]

$41''$. For the light of Jupiter is appreciably spread by unequal refrangibility, and this spreading has a lesser ratio to Jupiter's diameter in longer and more perfect telescopes than in shorter and less perfect ones. The times in which two satellites, the first and third, passed across the body of Jupiter, from the beginning of ingress to the beginning of egress, and from the completion of ingress to the completion of egress, were observed with the help of the same longer telescope. And the diameter of Jupiter at its mean distance from earth came out to be $37\frac{1}{8}''$ by the transit of the first satellite, and $37\frac{3}{8}''$ by the transit of the third. Further, the time in which the shadow of the first satellite passed across the body of Jupiter was observed, and from this the diameter of Jupiter at its mean distance from earth came out to be about $37''$. Let us assume that its diameter is about $37\frac{1}{4}''$ very nearly, and the greatest elongations of the first, second, third, and fourth satellites will be equal to 5.965, 9.494, 15.141, and 26.63 semidiameters of Jupiter, respectively.

Commentary on Phenomenon 1

As the title "Phenomena" of this general section indicates, its goal is to lay out a few carefully selected phenomena evident in our universe. After this has been accomplished, Newton will proceed to show that these phenomena, together with the theoretical structures he erected in Book I, can establish general principles, which, in turn, can account for a wide variety of other phenomena. Because Newton had formulated a variety of theoretical structures in his Book I, e.g., various force fields, the phenomena presented in Book III will allow readers to determine which of the possible universes is their universe.

In "Phenomenon 1," Newton draws on observations of the four satellites of Jupiter, which objects Galileo had discovered, to show that these four bodies obey Kepler's second and third laws. In setting out his third law, Kepler had not applied it to the planetary satellites, our Moon and the four Galilean satellites being the only satellites known at that time. Nonetheless, Newton realized that Kepler's third law should apply to the satellites of any object, with the qualification that the constant would be different for different systems. As one can see, the level of agreement Newton found is impressive. The only mathematical point that may need commentary is that it is important to note that Kepler's claim—that for $T =$ an object's orbital period and $D =$ the distance of the object around which it moves, then $T^2 \propto D^3$—is mathematically equivalent to the form that Newton gives it: $T \propto D^{3/2}$.

Newton then (see below) extends his analysis by examining five satellites of Saturn, which five objects had all been discovered in the period between 1655, when Huygens discovered Titan, the largest Saturnian satellite, and 1684, when Jacques Cassini of the Paris Observatory discovered Dione and Tethys.

Phenomena, continued[50]

Phenomenon 2

That the planets around Saturn, by radii drawn to Saturn, describe areas proportional to the times, and that their periodic times, the fixed stars being at rest, are in the sesquiplicate ratio of the distances from Saturn's center.

Cassini expressly stated their distances from the center of Saturn and their periodic times, from his observations, as follows.

Periodic Times of the Saturnian Satellites

1d 21h 18m 27s	2d 17h 41m 22s	4d 12h 25m 12s	15d 22h 41m 14s	79d 7h 48m 0s

Distances of the Satellites from the Center of Saturn in Semidiameters of the Ring

From the Observations	$1\frac{19}{20}$	$2\frac{1}{2}$	$3\frac{1}{2}$	8	24
From the Periodic Times	1.93	2.47	3.45	8	23.35

The greatest elongation of the fourth satellite from the center of Saturn is usually reckoned from observations to be about eight semidiameters, approximately. However, the greatest elongation of this satellite from the center of Saturn, taken with the best micrometer in Huygens's 123-foot-long telescope, came out to be eight semidiameters plus seven tenths of a semidiameter. And from this observation and the periodic times, the distances of the satellites from the center of Saturn, in semidiameters of the ring, are 2.1, 2.69, 3.75, 8.7, & 25.35. In the same telescope, the diameter of Saturn was to the diameter of the ring as 3 to 7, and the diameter of the ring, on the 28th and 29th of May, 1719, came out to be 43″. And hence the diameter of the ring at Saturn's mean distance from the earth is 42″, and Saturn's diameter is 18″. These things are so in the longest and best telescopes, because the apparent magnitudes of the celestial bodies in longer telescopes have a greater ratio to the spreading out of light at the edges of those bodies than they do in shorter telescopes. If all spurious light be disregarded, the remaining diameter of Saturn will be hardly greater than 16″.

---

[50] Densmore, *Central Argument* (footnote 19), pp. 321–322, 325, and 334–336.

Phenomenon 3

The five primary planets, Mercury, Venus, Mars, Jupiter, and Saturn, enclose the sun in their orbits.

That Mercury and Venus revolve around the sun is demonstrated from their lunar phases. When they shine with a full face, they are located beyond the sun, when halved they are even with the sun, and when sickle-shaped they are this side of the sun, sometimes passing across its disk in the manner of sunspots. Further, from Mars's full face near conjunction with the sun, and its gibbous face in the quadratures, it is certain that it encompasses the sun. The same is also demonstrated of Jupiter and Saturn from their ever full phases, for it is manifest from the shadows of satellites cast upon them that these shine with light borrowed from the sun.

Phenomenon 4

That the periodic times of the five primary planets, and that of the sun around the earth or of the earth around the sun (the fixed stars being at rest), are in the sesquiplicate ratio of the mean distances from the sun.

This ratio discovered by Kepler is acknowledged by all. The periodic times are necessarily the same, as well as the dimensions of the orbits, whether the sun revolve around the earth or the earth around the sun. And concerning the measure of the periodic times, there is unanimity among all astronomers. But Kepler and Bullialdus have determined the magnitudes of the orbs from the observations most diligently of all, and the mean distances which correspond to the periodic times, do not differ perceptibly from the distances which they have found, and for the most part fall in between them, as may be seen in the following table.

The Periodic Times of the Planets and the Earth about the Sun
With Respect to the Fixed Stars, in Days and Decimal Parts of a Day

Saturn	Jupiter	Mars	Earth	Venus	Mercury
10,759.275	4,332.514	686.9785	365.2565	224.6176	87.9692

Mean Distances of the Planets and the Earth from the Sun

	According to Kepler	According to Bullialdus	According to the Periodic Times
Saturn	951,000	954,198	954,006
Jupiter	519,650	522,520	520,096
Mars	152,350	152,350	152,369
Earth	100,000	100,000	100,000
Venus	72,400	72,398	72,333
Mercury	38,806	38,585	38,710

Concerning the distances of Mercury and Venus from the sun there is no room for dispute, since these are determined by their elongations from the sun. Further, concerning the distances of the superior planets from the sun, all dispute is removed by eclipses of Jupiter's satellites. For by those eclipses is determined the position of the shadow that Jupiter casts, and thereby is obtained Jupiter's heliocentric longitude. But from the heliocentric and geocentric longitudes compared with each other, the distance of Jupiter is determined.

Commentary on Phenomenon 4

It is interesting that at this point, particularly in his statement of Phenomenon 4, Newton as yet leaves open the dispute between the Copernican and the Tychonic systems. It is clear, however, from the caption for the table given in his discussion of Phenomenon 4 that he favors the former. As we shall see, Newton will eventually need to argue for the Copernican system (with an important modification) because his own system would have been impossible without it. On the other hand, in a certain sense, it was the Newtonian system that showed how the Copernican system could be true.

Phenomena, continued[51]

Phenomenon 5

That the primary planets, by radii drawn to the earth, describe areas by no means proportional to the times, but by radii drawn to the sun, do traverse areas proportional to the times.

For with respect to the earth, they now progress, now are stationary, now even retrogress; but with respect to the sun, they always progress, and do so with a motion very nearly uniform, but a little faster at perihelia and slower at aphelia, so as to make the description of areas equable. The proposition is very well known to astronomers, and in Jupiter particularly it is demonstrated by eclipses of the satellites, by which eclipses we have said that the heliocentric longitudes of this planet and its distances from the sun are determined.

---

[51] Densmore, *Central Argument* (footnote 19), pp. 347 and 349.

Phenomenon 6

That the moon by a radius drawn to the center of the earth describes an area proportional to the time.

This is clear from a comparison of the moon's apparent motion with its apparent diameter. However, the lunar motion is perturbed somewhat by the sun's force, but in these phenomena I neglect minute, imperceptible errors.

Propositions[52]

Proposition 1

That the forces by which the planets around Jupiter are perpetually drawn back from rectilinear motions and held back in their orbits, look to the center of Jupiter, and are inversely as the squares of the distances from the same center.

The former part of the proposition is clear from the first Phenomenon and the second or third Proposition of the first Book, and the latter part is clear from the first Phenomenon and the sixth Corollary of the fourth Proposition of the same Book.

The same is understood of the planets that accompany Saturn, from the second Phenomenon.

Proposition 2

That the forces by which the primary planets are perpetually drawn back from rectilinear motions, and are held back in their orbits, look to the sun, and are inversely as the squares of the distances from its center.

The former part of the proposition is clear from the fifth Phenomenon and the second Proposition of the first Book, and the latter part is clear from the fourth Phenomenon and the fourth Proposition of the same Book. However, this part of the Proposition is demonstrated with greatest accuracy by the aphelia being at rest. For (by Corollary 1 Proposition 45 Book I) the least departure from the duplicate ratio would be bound to effect a noticeable motion of the apsides in single revolutions, and an enormous one in many revolutions.

---

[52] Densmore, *Central Argument* (footnote 19), pp. 351, 353, 356, and 360–361.

Proposition 3

That the force by which the moon is held back in its orbit looks to the earth, and is inversely as the square of the distance of places from its center.

The former part of the assertion is clear from the sixth Phenomenon and the second or third Proposition of the first Book, and the latter part by the very slow motion of the lunar apogee. For that motion, which in single revolutions is but three degrees and three minutes forward, can be disregarded. For it is clear (by Corollary 1 Proposition 45 Book I) that if the distance of the moon from the center of the earth be called D,[53] the force by which such a motion would arise would be inversely as $D^{2\frac{4}{253}}$, that is, inversely as that power of D whose index is $2\frac{4}{253}$, or in other words, in a ratio of the distance a little greater than the duplicate inversely, but which approaches closer by $59\frac{3}{4}$ parts to the duplicate than to the triplicate. It arises, moreover, by the action of the sun (as will be said below), and for that reason is to be neglected here. It therefore remains that that force which looks to the earth is inversely as D^2.[54] A fact which will also be more fully established by comparing this force with the force of gravity, as it is in the following Proposition.

Corollary

If the mean centripetal force by which the moon is kept in its orbit be increased, first in the ratio of 177 to 178, and then in the duplicate ratio of the semidiameter of the earth to the mean distance of the moon's center from the earth's center, the centripetal force on the moon at the surface of the earth will be obtained, it being supposed that this force always increases in the inverse duplicate ratio of the altitude in descending to the surface of the earth.

Proposition 4

That the moon gravitates towards the earth, and is always drawn back from rectilinear motion, and held back in its orbit, by the force of gravity.

---

[53]The translation follows the first edition here. Later editions needlessly complicate the definition of D. [translator's note]

[54]For this sentence, the translation follows the first edition. In later editions, Newton replaced with a recomputation based upon other data regarding the Sun's disturbing force on the Moon. Since the computation is difficult and does not add materially to the argument, it is omitted. [translator's note]

The moon's mean distance from the earth in the syzygies,[55] in terrestrial semidiameters, is 59 according to Ptolemy and most astronomers; 60 according to Wendelin and Huygens, $60\frac{1}{3}$ according to Copernicus, $60\frac{2}{5}$ according to Streete, and $56\frac{1}{2}$ according to Tycho. But Tycho and those who follow his tables of refraction in setting a greater refraction—by as much as four or five minutes—for the sun and moon than for the fixed stars (in complete opposition to the nature of light), had increased the parallax of the moon by the same number of minutes; that is, by as much as the twelfth or fifteenth part of the whole parallax. Let this error be corrected, and the distance will come out to be $60\frac{1}{2}$ terrestrial semidiameters, more or less, about what was assigned by the others. Let us assume that the mean distance is sixty semidiameters at the syzygies, and that the lunar period with respect to the fixed stars amounts to 27 days, 7 hours, and 43 minutes, as is stated by the astronomers; and that the circumference of the earth is 123,249,600 Paris feet, as is established by the measuring Frenchmen. If the moon be supposed to be deprived of all motion and dropped, so as to descend towards the earth, under the influence of all that force by which (by Proposition 3 Corollary) it is held back in its orbit, it will in falling traverse $15\frac{1}{12}$ Paris feet in the space of one minute. This conclusion comes from a computation based either upon Proposition 36 of the first Book or (what amounts to the same thing) the ninth Corollary of the fourth Proposition of the same Book. For the versed sine of that arc which the moon in its mean motion describes in the time of one minute at a distance of sixty terrestrial semidiameters, is about $15\frac{1}{12}$ Paris feet, or more accurately, 15 feet 1 inch and $1\frac{4}{9}$ lines. Whence, since in approaching the earth that force increases in the inverse of the duplicate ratio of the distance, and is thus greater at the surface of the earth by 60×60 parts than at the moon, a body, in falling by that force in our regions, ought to describe a space of $60 \times 60 \times 15\frac{1}{12}$ Paris feet in the space of one minute, and in the space of one second, $15\frac{1}{12}$ feet, or more accurately, 15 feet 1 inch and $1\frac{4}{9}$ lines. And heavy bodies on earth do in fact descend with the same force. For the length of a pendulum oscillating in seconds, at the latitude of Paris, is three Paris feet $8\frac{1}{2}$ lines, as Huygens has observed. And the height which a heavy body traverses in falling in the time of one second, is to half the length of this pendulum, in the duplicate ratio of the circumference of the circle to its diameter (as Huygens has also pointed out). It is therefore 15 Paris feet 1 inch $1\frac{7}{9}$ lines. And because the force which

---

[55] The syzygy points for the Moon are the two points in the Moon's orbit when the moon lies on the line extending through both the Earth and the Sun. [MJC]

holds the moon back in its orbit, if it should descend to the surface of the earth, comes out equal to our force of gravity, therefore (by Rules 1 and 2) it is that very force which we are accustomed to call gravity. For if gravity were different from it, bodies, in seeking the earth with the two forces conjoined, would descend twice as fast, and in falling in the space of one second would describe $30\frac{1}{6}$ Paris feet, in complete opposition to experience.

This computation is based upon the hypothesis that the earth is at rest. For if the earth and the moon should move around the sun, and should also at the same time move around their common center of gravity, the law of gravity remaining the same, the distance of the centers of the moon and the earth from each other will be about $60\frac{1}{2}$ terrestrial semidiameters, as will be clear to anyone undertaking the computation. And the computation can be undertaken by Proposition 60 of Book I.

Commentary on Proposition 4

What Newton has done in his discussion of his Proposition 4 is basically what was described earlier in this chapter under the heading "Application of the Law of Centripetal Acceleration to the Moon" (pp. 123–125).

Propositions, continued[56]

Scholium to Proposition III.4

The demonstration of the proposition can be more amply displayed thus. If many moons were to revolve around the earth, exactly as in the system of Saturn or Jupiter, their periodic times (by an argument of induction) would observe the law of the planets discovered by Kepler, and therefore, their centripetal forces would be inversely as the squares of the distances from the center of the earth, by Proposition 1 of this Book. And if the lowest of these were small, and were nearly to touch the peaks of the highest mountains, its centripetal force, which keeps it in its orbit, would (by the foregoing computation) be very nearly equal to the gravities of bodies on the peaks of those mountains. It would thus come to pass that if the same small moon were deprived of all the motion by which it proceeds in its orbit, it would

---

[56] Densmore, *Central Argument* (footnote 19), pp. 376–377, 380–381, 383, 402–404, 419–421, 427, and 460–461.

descend to earth because of the loss of the centrifugal force by which it had remained in its orbit, and would do this with the same velocity with which heavy bodies fall on the peaks of those mountains, because of the equality of the forces with which they descend. And if that force by which that small lowest moon descends were different from gravity, and that small moon were also heavy towards the earth in the manner of the bodies on the peaks of the mountains, the same small moon under the two conjoined forces would descend twice as fast. Therefore, since both forces—the latter ones, of the heavy bodies, and the former ones, of the moons—look to the center of the earth, and are similar and equal to each other, these same [forces] (by Rules 1 and 2) will have the same cause. And consequently, that force by which the moon is kept back in its orbit, will be that very force which we usually call "gravity," and this must above all be true lest the small moon at the peak of the mountain either lack gravity, or fall twice as fast as heavy bodies usually fall.

Proposition 5

That the planets around Jupiter gravitate towards Jupiter, the planets around Saturn towards Saturn, and the planets around the sun towards the sun, and by the force of their gravity are always drawn back from rectilinear motions and confined to curvilinear orbits.

For the revolutions of the planets around Jupiter about Jupiter, of the planets around Saturn about Saturn, and of Mercury and Venus and the rest of the planets around the sun about the sun, are phenomena of the same kind as the revolution of the moon about the earth, and for that reason (by Rule 2) depend upon causes of the same kind. This is primarily because it is demonstrated that the forces upon which those revolutions depend, look to the centers of Jupiter, Saturn, and the sun, and in receding from Jupiter, Saturn, and the sun, decrease in the same ratio and by the [same] law by which the force of gravity decreases in recession from earth.

III.5 Corollary 1

Gravity is therefore given towards all planets without exception. For no one doubts that Venus, Mercury, and the rest are bodies of the same kind as Jupiter and Saturn. And since by the Third Law of Motion all attraction is mutual, Jupiter will gravitate towards all of its satellites, Saturn towards all of its [satellites], the earth towards the moon, and the sun towards all the primary planets.

III.5 Corollary 2

Gravity, which looks to each planet, is inversely as the square of the distances of places from its center.

III.5 Corollary 3

All planets are mutually heavy towards each other by Corollaries 1 and 2. And hence, Jupiter and Saturn, attracting each other near conjunction, perturb each others' motions perceptibly; the sun perturbs the lunar motions; the sun and moon perturb our sea; as will be explained in what follows.

Scholium

Hitherto we have called that force by which the celestial bodies are confined to their orbits "centripetal." It is now established that the same [force] is gravity, and for that reason we shall hereafter call it "gravity." For the cause of that centripetal force by which the moon is confined to [its] orb is obligated to extend to all planets, by Rules 1, 2, and 4.

Proposition 6

That all bodies gravitate to each of the planets, and that their weights towards whichever particular planet you please, at equal distances from the center of the planet, are proportional to the quantity of matter in each.

That the descent of all heavy bodies to earth (if the unequal retardation arising from the air's very slight resistance be subtracted) occurs in equal times, others have observed for some time now, and the equality of times can be noted very accurately indeed in pendulums. I have made a trial with gold, silver, lead, glass, sand, common salt, wood, water, [and] wheat. I prepared two equal and round wooden boxes. I filled one with wood, and suspended the same weight of gold at the center of oscillation of the other (as exactly as I could). When the boxes were hung from equal threads of eleven feet, they constituted pendulums entirely similar as to weight, shape, and air resistance, and, placed side by side, they went to and fro together for a very long time with equal oscillations. Accordingly, the amount of matter in the gold (by Book II Proposition 24 Corollaries 1 and 6) was to the amount of matter in the wood as the action of the motive force on all the gold was to the action of the same on all the wood, that is, as weight to weight.[57] And thus it was in the others. In bodies of the same weight, a difference in matter of no greater than the thousandth part of the whole matter could have been plainly perceived in these experiments.

But now there is no doubt that the nature of gravity in the planets is the same as on earth. For let these terrestrial bodies be imagined to be raised right up to the orbit of the moon and to be dropped along with the moon, deprived of all motion, so as to fall to earth at the same time. Through what has already been shown, it is certain that in equal times they will describe equal distances with the moon, and therefore that they are to the quantity of matter in the moon as their weights are to its weight. Further, since the satellites revolve in times that are in the sesquiplicate ratio of the distances from the center of Jupiter, their accelerative gravities towards Jupiter will be inversely as the squares of the distances from the center of Jupiter, and consequently, at equal distances from Jupiter their accelerative gravities will come out equal. Further, in falling from equal altitudes in equal times, they would describe equal spaces, exactly as happens with heavy bodies in this our earth. And by the same argument, the planets around the sun, dropped from equal distances from the sun, would in their descent to the sun describe equal spaces in equal times.

But the forces by which unequal bodies are equally accelerated are as the bodies; that is, the weights are as the quantities of matter in the planets. Further, that the weights of Jupiter and its satellites towards the sun are proportional to the quantities of their matter is evident from the motion of the satellites, as regular as can be, by Book I Proposition 65 Corollary 3. For if some of them were to be pulled towards the sun more in proportion to their quantity of matter than the others, the motions of the satellites (by Book 1 Proposition 65 Corollary 2) would be perturbed by the inequality of the attraction. If, at equal distances from the sun, some satellite were heavier towards the sun in proportion to its quantity of matter than Jupiter in proportion to its quantity of matter, in any given ratio (say, d to e), the distance between the center of the sun and the center of the satellite's orbit would always be greater than the distance between the center of the sun and the center of Jupiter in the subduplicate ratio, approximately, as I have

---

[57] The point of this experiment is to show the identity, at least in regard to quantitative measure, of gravitational and inertial mass. [MJC]

found by entering upon a certain computation. And if the satellite were less heavy towards the sun in that ratio d to e, the distance of the center of the satellite's orbit from the sun would be less than the distance of Jupiter's center from the sun in that subduplicate ratio. And thus if at equal distances from the sun, the accelerative gravity of any satellite you please towards the sun be greater or less than the accelerative gravity of Jupiter towards the sun, by only the thousandth part of the whole gravity, the distance of the center of the satellite's orbit from the sun would be greater or less than the distance of Jupiter from the sun by the $\frac{1}{2000}$ part of the whole distance, that is, by the fifth part of the outermost satellite's distance from the center of Jupiter—an eccentricity of the orbit that would be very easily perceptible. But the orbits of the satellites are concentric with Jupiter, and consequently the accelerative gravities of Jupiter and of the satellites towards the sun are equal among themselves. And by the same argument, the weights of Saturn and its companions towards the sun, at equal distances from the sun, are as the quantities of matter in them, and the weights of the moon and the earth towards the sun are either nothing, or accurately proportional to their masses. But they are something, by Proposition 5 Corollaries 1 and 3.

And furthermore, the weights of individual parts of any planet to any other whatever are to each other as the matter in the individual parts. For if certain parts were to gravitate more, others less, than according to the quantity of matter, the whole planet, according to the kind of parts with which it mostly abounds, would gravitate more or less than according to the quantity of matter of the whole. Nor does it matter whether those parts be external or internal. For if, for example, the terrestrial bodies which are with us were imagined to be raised to the orbit of the moon, and compared with the body of the moon, if their weights were to the weights of the external parts of the moon as the quantities of matter in the same bodies, but were to the weights of the internal parts in a greater or less ratio, the same [weights] would be to the weight of the whole moon in a greater or less ratio, contrary to what was shown above.

III.6 Corollary 1

Hence the weights of bodies do not depend upon their shapes and textures. For if the weights could vary with shapes, they would be greater or less, according to the variety of shapes, in an equal matter, entirely contrary to experience.

III.6 Corollary 2

All bodies that are around the earth are heavy towards the earth, and the weights of all that stand at the same distance from the center of the earth are as the quantities of matter in them. This is a quality of all [bodies] upon which experiments can be carried out, and accordingly must be affirmed universally, by Rule 3. If the aether or any other body whatever either were entirely lacking in gravity, or were to gravitate less in proportion to its quantity of matter, then, because (in the opinion of Aristotle, Descartes, and others) it does not differ from other bodies except in the shape of matter, the same [body], by a gradual alteration of shape, could be transformed into a body of the same condition as those that gravitate most in proportion to the quantity of matter. And in turn, the heaviest bodies, by gradually assuming the shape of the former [body], would be able gradually to shed their gravity. And thus the weights would depend upon the shapes of bodies, and would be capable of changing with the shapes, contrary to what is proved in the preceding corollary.

III.6 Corollary 3

All spaces are not equally full. For if all spaces were equally full, the specific gravity of the fluid that fills the region of the air would concede nothing to the specific gravity of quicksilver or gold or any other extremely dense body, because the matter would be of the greatest density; and consequently, neither gold nor any other body whatever could descend in air. For unless bodies in fluids are specifically heavier, they do not descend in the least. And if the quantity of matter in a given space could be diminished somewhat through rarefaction, why might it not be diminished *in infinitum*?

III.6 Corollary 4

If all the solid particles of all bodies be of the same density, and not be capable of being rarefied without pores, the vacuum is given. I declare to be of the same density, those things whose forces of inertia are as [their] magnitudes.

III.6 Corollary 5

The force of gravity is different in kind from the magnetic force. For magnetic attraction is not as the matter attracted. Some bodies are pulled more, others less, most are not pulled. And the magnetic force in one and the same body can be intensified and remitted, and is not infrequently far greater in proportion to the quantity of matter than the force of gravity, and in withdrawal from the magnet it decreases, not in the duplicate, but nearly in the triplicate ratio of the distance, as far as I have been able to ascertain by some rather crude observations.

Proposition 7

That gravity is given towards bodies universally, and that it is proportional to the quantity of matter in each.

That all the planets are mutually heavy towards each other, we have now already proved, as well as that gravity towards any one of them, considered separately, is inversely as the square of the distance of places from the center of the planet. And the consequence of this (by Book I Proposition 69 and its corollaries) is that gravity towards all is proportional to the matter in the same bodies.

Further, since all the parts of any planet you please *A* are heavy towards any planet you please *B*, and the gravity of any part is to the gravity of the whole as the matter of the part is to the matter of the whole, and to every action there is an equal reaction (by the Third Law of Motion), [therefore] the planet *B* will in turn gravitate towards all the parts of the planet *A*, and its gravity towards any particular part will be to its gravity towards the whole as the matter of the part to the matter of the whole.   Q.E.D.

III.7 Corollary 1[58]

Gravity towards the whole planet therefore arises from, and is compounded of, the gravity towards its individual parts. We have an example of this in magnetic and electric attractions. For all the attraction towards the whole arises from the attractions towards the individual parts. This will be understood in regards to gravity by conceiving of several smaller planets coming together into one globe and forming a larger planet. For it will be necessary that the force of the whole arise from the forces of the component parts. If one should object that by this law all bodies that are around us would be required to gravitate mutually towards each other, although gravity of this kind is by no means felt, I answer that, because gravity towards these bodies

---

[58]William H. Donahue prepared the translation of this and the next corollary specifically for this volume. [MJC]

is to gravity towards the earth as these bodies are to the whole earth, the gravity towards these bodies is far less than what can be perceived.

III.7 Corollary 2

Gravity towards the individual equal parts of a body is inversely as the square of the distance of places from the particles. This is evident from Book I Prop. 74 Cor. 3.

Proposition 8

If the matter of two globes mutually gravitating to each other be everywhere homogeneous in regions that are equidistant from the centers, the weight of each globe towards the other will be inversely as the square of the distance between the centers.

After I had found that gravity in an entire planet arises and is compounded from the gravities to the parts, and that in the individual parts it is inversely proportional to the squares of the distances from the parts, I was in doubt whether that inverse duplicate proportion might apply accurately in the whole force compounded of many forces, or only approximately. For it might happen that the proportion which applies accurately enough at greater distances, would err noticeably near the surface of the planet because of the unequal distances and dissimilar positions of the particles. But at last, by Propositions 75 and 76 of Book I and their corollaries, I understood the truth of the proposition that is here in question.

Commentary on Book III from Phenomenon 1 to Proposition 8

These materials are relatively straightforward; to follow them in every detail would be difficult, but to see their overall significance is relatively easy. Gradually and carefully Newton has built his case for the most extraordinary result established in his mechanics: the law of universal gravitation. For example, in Propositions 6 and 7 he shows how the masses of objects are involved and in Proposition 8 he draws on Propositions 71–76 of Book I to deal with the question of whether the attraction of a sphere can be treated as if all its mass were concentrated at the center, including even the special case where the attracted body is near the surface of the attracting sphere. Thus Newton has arrived at what modern treatises on mechanics express in the equation

$$F = \frac{GM_1M_2}{D^2}.$$

In this equation, F is the gravitational force of attraction between two masses M_1 and M_2 separated by the distance D between their centers, and G is a constant of proportionality called the gravitational constant. Newton expressed each of the fundamental relationships in this equation individually, for example, that the force varies inversely with the square of the distance separating the centers of the attracting bodies. Because Newton expressed these relationships as proportions and not as an equation, he had no need of a constant of proportionality, so G does not appear in his *Principia*. G was carefully measured in the late eighteenth century by Nevil Maskelyne and also by Henry Cavendish. Its current value is $6.67 \cdot 10^{-11}$ newtons $\cdot \frac{\text{meters}^2}{\text{kilograms}^2}$. With this powerful formula in hand, let us proceed to a few illustrative problems.

Problems

1. Suppose the Moon were replaced in its orbit by a one pound sphere of cheese. Let the latter start off with the position and velocity that the Moon has had. Determine its subsequent path.

Solution:
Let us combine Newton's second law of motion with his law of gravitation, using M_e for the mass of the Earth and M_c for the mass of the cheese: $F = M_c a = \frac{G M_c M_e}{D^2}$. Dividing through by M_c, we find that the acceleration of the chunk of cheese will be: $a = \frac{G M_e}{D^2}$. Note that M_c does not appear in this equation. That mathematical fact implies that the acceleration of the object does not depend on the object's mass. That is to say that its acceleration and consequently its path will be exactly that of the Moon. This problem, you should realize, is nothing more than a variation of the leaning tower of Pisa experiment!

2. You and your date stand three meters apart. One of you weighs 800 newtons, the other 650 newtons. Calculate the force of the attraction (gravitational) between the two of you.

Solution:
The masses of the two persons are $\frac{800}{9.8} = 81.7$ kilograms and $\frac{650}{9.8} = 66.3$ kilograms. Substituting these two values into the law of gravitation and using the figure for G given previously, we have:

$$F = 6.67 \cdot 10^{-11} \text{ newtons} \cdot \frac{\text{meters}^2}{\text{kilograms}^2} \cdot \frac{81.7 \text{ kg} \cdot 66.3 \text{ kg}}{3^2 \text{ meters}^2} = 40.1 \cdot 10^{-9} \text{ newtons.}$$

3. Weigh Saturn, comparing its mass to that of the Earth. In the calculation,

use the facts that Saturn's chief moon, Titan, has a period of 16 days and is 760,000 miles from Saturn.

Solution:

Applying the law of gravitation and Newton's second law to both the Earth-Luna (Moon) and Saturn-Titan systems gives

$$F_{L \Leftrightarrow E} = M_L a_L = \frac{GM_L M_E}{D_{L \Leftrightarrow E}^2} \text{ and } F_{T \Leftrightarrow S} = M_T a_T = \frac{GM_T M_S}{D_{T \Leftrightarrow S}^2}.$$

After dividing the first equation by M_L and the second by M_T, we can divide the equation at the left by the one at the right to get:

$$\frac{GM_E / D_{L \Leftrightarrow E}^2}{GM_S / D_{T \Leftrightarrow S}^2} = \frac{a_L}{a_T}.$$

This simplifies to $\dfrac{M_E}{M_S} = \dfrac{a_L}{a_T} \dfrac{D_{L \Leftrightarrow E}^2}{D_{T \Leftrightarrow S}^2}.$

To get a_L and a_T, we use the formula for centripetal acceleration:

$$a = v^2/r = \left(\frac{2\pi r}{T} \right)^2 \bigg/ r = \frac{(2\pi)^2 r}{T^2}.$$

Applying this to the earlier equation, we have:

$$\frac{M_E}{M_S} = \frac{(2\pi)^2 \cdot (D_{L \Leftrightarrow E} / T_L^2) \cdot (D_{L \Leftrightarrow E})^2}{(2\pi)^2 \cdot (D_{T \Leftrightarrow S} / T_T^2) \cdot (D_{T \Leftrightarrow S})^2}. \text{ Hence } \frac{M_E}{M_S} = \frac{(T_T)^2 \cdot (D_{L \Leftrightarrow E})^3}{(T_L)^2 \cdot (D_{T \Leftrightarrow S})^3}.$$

Substituting our data into this equation gives:

$$\frac{M_E}{M_S} = \frac{(16 \text{ days})^2 \cdot (240,000 \text{ miles})^3}{(27.3 \text{ days})^2 \cdot (760,000 \text{ miles})^3} = .0108 = \frac{1}{93}.$$

We have thus determined that Saturn is about 93 times more massive than the Earth. This method can be used to determine the relative masses of any two planets, provided that each has at least one satellite orbiting it. This illustration is directly relevant to Newton's *Principia*; as you will see in Corollary II to Proposition VIII, Newton carried out such a calculation, determining by this means the relative masses of the Sun, Earth, Jupiter, and Saturn.

Propositions, continued[59]

III.8 Corollary 1

Hence can be found and compared among each other the weights of bodies on different planets. For the weights of equal bodies revolving in circles about the planets are (by Book I Proposition 4 Corollary 2) as the diameters of the circles directly and the squares of the periodic times inversely; and the weights at the surfaces of the planets, or at any other distances from the planets you please, are greater or less (by this proposition) inversely in the duplicate ratio of the distances. Thus from the periodic times of Venus around the sun, 224 days and $16\frac{3}{4}$ hours, of the outermost satellite of Jupiter around Jupiter, 16 days and $16\frac{8}{15}$ hours, of the Huygenian satellite around Saturn, 15 days and $22\frac{2}{3}$ hours, and of the moon around the earth, 27 days 7 hours 43 minutes, compared with the mean distance of Venus from the sun and with the greatest heliocentric elongations of the outermost of Jupiter's satellites from the center of Jupiter, $8'\ 16''$, of the Huygenian satellite from the center of Saturn, $3'\ 4''$, and of the moon from the center of the earth, $10'\ 33''$, by entering into a computation I have found that the weights of equal bodies at equal distances from the center of the sun, Jupiter, Saturn, and earth, to the sun, Jupiter, Saturn, and earth, respectively, are as 1, 1/1067, 1/3021, and 1/169,282 respectively, and when the distances are increased or decreased, the weights are decreased or increased in the duplicate ratio. Thus the weights of equal bodies on the sun, Jupiter, Saturn, and earth, at distances of 10000, 997, 791, and 109 from their centers, and accordingly on their surfaces, will be as 10000, 943, 529, and 435, respectively. The magnitude of weights of bodies on the surface of the moon will be told in what follows.

III.8 Corollary 2

The quantity of matter in the individual planets is also found. For the quantities of matter in the planets are as their forces at equal distances from their centers; that is, in the sun, Jupiter, Saturn, and the earth, they are as 1, 1/1067, 1/3021, and 1/169,282 respectively. If the sun's parallax be set at greater or less than $10''\ 30'''$, the quantity of matter in the earth will have to be increased or diminished in the triplicate ratio.

---

[59] Densmore, *Central Argument* (footnote 19), pp. 465–466 and 476.

III.8 Corollary 3[60]

The densities of the planets also become known. For, by Book I Prop. 72, at the surfaces of homogeneous spheres, the weights of equal and homogeneous bodies towards the spheres are as the diameters of the spheres; therefore, the densities of the homogeneous spheres are as those weights divided by the diameters of the spheres. Now the true diameters of the sun, Jupiter, Saturn, and earth were found to be to each other as 10,000, 997, 791, and 109, and the weights towards the same bodies are as 10,000, 943, 529, and 435, respectively, and therefore the densities are as 100, 94, 67, and 400. The density of the earth that results from this computation does not depend on the sun's parallax, but is determined by the moon's parallax, and accordingly is determined correctly here. The sun is therefore a little denser than Jupiter, and Jupiter a little denser than Saturn, and the earth is four times as dense as the sun. For the sun becomes rarefied through its tremendous heat. The moon, on the other hand, is denser than the earth, as will be made clear in what follows.

III.8 Corollary 4

Therefore, the smaller planets are more dense, other things being equal. For the force of gravity at their surfaces thus approaches nearer to equality. But the planets that are nearer the sun, other things being equal, are also denser, as Jupiter is denser than Saturn, and the earth denser than Jupiter. No doubt the planets had to be placed at different distances from the sun so that each might benefit from the sun's heat to a greater or lesser degree according to the degree of its density. Our water would become rigid if the earth had been placed at the orb of Saturn, and at the orb of Mercury it would immediately fly off as a vapor. For the sun's light, to which its heat is proportional, is seven times denser at the orb of Mercury than it is with us, and I have found by experiment with a thermometer that with a heat that is seven times that of the summer sun, water boils away. There is, however, no doubt that the matter of Mercury is accommodated to heat, and for that reason is denser than this matter of ours, since all denser matter requires greater heat to undergo natural processes.

---

[60]The translation of this and the next corollary were prepared by William H. Donahue for this volume. [MJC]

Commentary on Newton's Corollaries to Proposition 8

In his corollaries to Proposition 8 in Book III of his *Principia*, Newton derives various interesting results about a number of the planets, using the method that we have already seen of determining the relative masses of bodies around which other bodies move. This method works only for those solar system objects that have other bodies orbiting them. Thus the method allows Newton to compare the masses of the Sun, Earth, Jupiter, and Saturn, but he cannot by this method estimate the masses of Mercury, Venus, Mars, or the Moon, because they do not have satellites moving around them.

Newton's comparison of the masses of the Sun, Earth, Jupiter, and Saturn can be put in the following tabular form, in which the modern values are also provided. Newton assigns the Sun a mass of 1 and then gives the masses of the other objects in terms of this value.

Mass of	Newton's Value	Modern Value
Sun	1	1
Jupiter	1/1067	1/1047
Saturn	1/3021	1/3497
Earth	1/169,282	1/333,000

These values reveal, on the one hand, the immensity of the mass of the Sun and, on the other hand, the relative insignificance of the mass of the Earth. Using his data, Newton could have calculated that it would, for example, take 169,282 Earths to weigh as much, or exert as great an attraction, as the Sun. In fact, his value for the mass of the Earth compared to the Sun was about double what we now know it to be. Not only is the Earth a relative lightweight in comparison with the Sun, but also in relation to Jupiter and Saturn. Using modern values, we can calculate that it would take 318 Earths to make a body as massive as Jupiter, whereas the mass of 95 Earths would be the same as Saturn's.

Newton also compares the intensity of the gravitational force at the surface of these four objects. In Corollary 1, he gives the following information, norming his values on an object that would weigh 10,000 units on the Sun:

Solar System Body	Weight of Object at the Surface of That Body
Earth	435
Jupiter	943
Saturn	529
Sun	10,000

Additional interesting information is provided as to the densities of the various bodies. These are, of course, simply derived from his earlier values for the masses of the bodies combined with observations of their diameters, from which, after a determination of their distances have been made, their volumes can be calculated.

Solar System Body	Density Assigned by Newton
Sun	100
Jupiter	94.5
Saturn	67
Earth	400

These results are rather surprising. They indicate that the Earth is far more dense than the Sun, Jupiter, or Saturn. Having found the Earth so much more dense than the giant outer planets, Newton concludes in Corollary 4 that as one gets nearer the Sun, the planets tend to be more dense. This generalization can be checked by a comparison with modern values:

Planet	Density (Modern Values in grams/cm^3)
Mercury	5.1
Venus	5.3
Earth	5.52
Mars	3.94
Jupiter	1.33
Saturn	.69

Another point worth noting is that Newton in his discussion of Corollary 2 states that the Sun has a parallax of 10.5″. This is a crucial value for astronomical calculations in that from it the actual astronomical distances are calculated. In fact, the distance from the Earth to the Sun is called the **astronomical unit** (abbreviated **a.u.**), a term that indicates its great significance. One example of this significance is that using the determination of the astronomical unit in conjunction with Kepler's third law, the distances in miles of all the planets from the Sun can be calculated. Of course, if the value of the angle of solar parallax, π, is off, then the derived distances will be off accordingly. We can determine the value that Newton had for the distance of the Earth from the Sun from his value for the solar parallax.

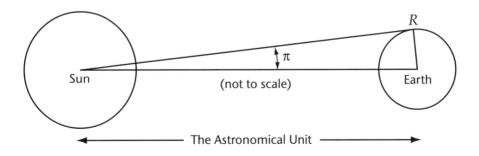

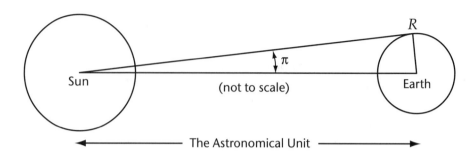

The Astronomical Unit

In making this calculation we need know only the radius of the Earth, R, and the angle of parallax, π. Newton's value for the radius of the Earth must not have been far off from the modern value, which is 3963 miles. Consequently, we have: $\tan \pi = \tan 10.5'' = 3963 \text{ miles}/X$, where X is the distance between the Sun and Earth. Substituting Newton's value for π gives:

$$\tan \pi = \tan 10.5'' = .0000509 = 3963/X,$$

which entails that

$$X = 78 \text{ million miles.}$$

From the facts that the modern value of the solar parallax is $8.7941''$ and the modern value for the distance of the Sun is about 93 million miles, one can see that Newton was substantially off, although far closer than Copernicus to the true distance of the Sun, the Copernican value being about 25 times smaller than the actual value.

These values may seem mere numbers that would have no religious or philosophical significance nor be of any interest to imaginative writers. That this was scarcely the case is shown by how these numbers impacted on ideas of extraterrestrials. For example, the huge mass assigned the Sun implied that because of the resultant immense gravitational attraction at the Sun, solarians would have to be extremely strong. Newton's figure in Corollary 4 showing that the rays of the Sun are seven times more intense on Mercury than on the Earth, not only deprived Mercury of water (as Newton noted), but also forced advocates of extraterrestrials to make Mercurians capable of withstanding very high temperatures. Early in the eighteenth century, the philosopher Christian Wolff calculated the height of Jupiterians from the fact that the Sun shines on Jupiter with far less intensity than on the Earth. Assuming that the size of the Jupiterians' eyes must be correspondingly much larger than ours to adjust to the diminished sunlight of that planet, Wolff estimated that Jupiterians must be about 13 feet tall. In the nineteenth century, William Whewell, no friend of extraterrestrials, created havoc for Jupiterians and Saturnians, or at least their terrestrial advocates, by pointing out that the low density of their planets—approximately that of water—entails that if Jupiterians and Saturnians exist, they would have to be fish-like in nature.

Propositions, continued[61]

Proposition 9

Gravity, when one goes downwards from the surfaces of planets, decreases in proportion to the distances from the center, very nearly.

If the matter of a planet were uniform as to density, this proposition would hold exactly, by Prop. 73 of Book I. Therefore, the error is as great as what could arise from a nonuniform density.

Proposition 10

The motions of the planets in the heavens can be maintained for a very long time.

In the Scholium to Book II Proposition 40, it was shown that a globe of frozen water, moving freely in our air, would give up 1/4586 part of its motion from the resistance of the air in traversing the length of its semidiameter. Moreover, the same ratio holds very nearly for globes of any size and speed. But now, the fact that the globe of our earth is denser than it would be if it were entirely made of water, I deduce as follows. If this globe were all made of water, whatever things were less dense than water would come out and float on top of it because of their lesser specific gravity. By this cause, an earthy globe, everywhere covered by the waters, would, if it were less dense than water, come out somewhere, and all the water would flow off of it and gather at the other side. And the account of our earth, surrounded on most sides by the seas, is similar to this. If this [earthy part] were not denser, it would come forth out of the seas, and a part of it proportional to its degree of lightness would protrude from the water, while all the seas gathered together on the other side. By the same argument, the sunspots are lighter than the sun's luminous matter upon which they float And in the formation of the planets, however that might be, at the time when the mass was fluid, all the heavier matter, leaving the water, sought out the center. Consequently, since the ordinary earth on top is about twice as heavy as water, and a little below, in mines, it

---

[61] The translations from Newton's *Principia* in this section with the exception of Proposition 13 were kindly provided for this volume by William H. Donahue. Donahue's translation of Proposition 13 appears in Dana Densmore, *Newton's* Principia: *The Central Argument*, with translations by William H. Donahue, pp. 477–478.

is found to be about three or four times, or even five times heavier, it is probable that the store of all the matter in the earth is about five or six times greater [in weight] than if it were all made of water, especially since it was just shown that the earth is about four times denser than Jupiter. As a result, if Jupiter be a little denser than water, it would, in the space of thirty days, in which it traverses the length of 459 of its semidiameters, give up almost one tenth part of its motion in a medium of the same density as our air. But at the same time, since the resistance of media decreases in the ratio of the weight and the density (so that water, which is $13\frac{3}{5}$ times lighter than quicksilver, resists less in the same ratio, and air, which is 860 times lighter than water, resists less in the same ratio), if an ascent be made to the heavens, where the weight of the medium in which the planets move is decreased into the immense space, the resistance will practically cease. In particular, we have shown in the Scholium to Book II Prop. 22 that if an ascent were made to an altitude of two hundred miles above the earth, the air there would be rarer than at the surface of the earth in the ratio of 30 to 0.0000000000003998, or about 75,000,000,000,000 to 1. And hence the star Jupiter, orbiting in a medium of the same density as that upper air, in the time of 1,000,000 years would not lose a tenth of a hundredth of a thousandth part [i.e., a millionth] of its motion from the resistance of the medium. At least for spaces near to the earth, nothing is found that would create resistance other than air, exhalations, and vapors. When these are exhausted thoroughly from a hollow glass cylinder, heavy bodies inside the glass fall freely and without any perceptible resistance: gold itself and the finest feather, dropped at the same time, fall with equal speed, and, traversing a height of four, six, or eight feet, hit the bottom at the same time, as has been established by experience. And for that reason, if an ascent be made to heavens void of air and exhalations, planets and comets will move without any perceptible resistance through those spaces for a very long time.

Hypothesis 1

The center of the system of the world is at rest.

This is granted by everyone, although some would argue that it is the earth, and others that it is the sun, that is at rest at the center of the system. Let us see what follows from this.

Proposition 11

The common center of gravity of the earth, the sun, and all the planets is at rest.

For (by Corollary 4 of the Laws) that center either is at rest or proceeds uniformly in a straight line. But by a perpetual proceeding of that center, the center of the world will also move, contrary to the hypothesis.

Proposition 12

The sun is driven about with a ceaseless motion, but never departs far from the common center of gravity of all the planets.

For, by Prop. 8 Cor. 2, the matter in the sun is the matter in Jupiter as 1,067 to 1, and Jupiter's distance from the sun is in a slightly greater ratio to the sun's diameter; therefore, the common center of gravity of Jupiter and the sun will fall at a point slightly above the sun's surface. By the same argument, the matter in the sun is to the matter in Saturn as 3,021 to 1, and Saturn's distance from the sun is the sun's semidiameter in a slightly smaller ratio; therefore, the common center of gravity of Saturn and the sun will fall at a point slightly beneath the sun's surface. And, treading in the footsteps of the same computation, if we placed the earth and all the planets on one side of the sun, the common center of gravity of all of them would be distant from the sun's center by barely one whole diameter of the sun. In other instances, the distance of the centers is always less. And therefore, since that center of gravity is always at rest, the sun will move in all directions, in accord with the various positions of the planets, but will never depart far from that center.

Corollary

Hence, the common center of gravity of the earth, the sun, and all the planets, is to be taken as the center of the world. For since the earth, the sun, and all the planets gravitate to each other mutually, and for that reason, following the laws of motion, are driven about ceaselessly, in accord with the force of their gravity, it is clear that the movable centers of these bodies cannot be taken as the unmoving center of the world. If the central location is to be given to that body to which all bodies gravitate most (as the opinion of most people holds), that privilege would have to be granted to the sun. But since the sun moves, an unmoving point will have to be chosen, from which the sun's center departs as little as possible, and from which that same center would move even less if the sun were denser and greater, so as to be moved less.

Proposition 13

The planets move in ellipses having their focus at the center of the sun, and by radii drawn to that center describe areas proportional to the times.

Hitherto we have reasoned about these motions from the phenomena. Now that the principles of the motions are known, we infer from these the celestial motions *a priori*. Because the weights of the planets towards the sun are inversely as the squares of the distances from the sun's center, if the sun were to be at rest and the remaining planets were not to act mutually upon each other, their orbits would come out elliptical, with the sun at the common focus. Also, the areas described would be proportional to the times (by Book I Proposition 1 and 11, and Proposition 13 Corollary 1). But the mutual actions of the planets upon each other are very slight (so that they can be disregarded), and (by Book I Proposition 66) they perturb the motions of the planets about a movable sun less than if those motions were carried out about a sun at rest.

Moreover, the action of Jupiter upon Saturn is not entirely to be disregarded. For the gravity to Jupiter is to the gravity to the sun (at equal distances) as 1 to 1,067, and therefore, at the conjunction of Jupiter and Saturn, since Saturn's distance from Jupiter is to Saturn's distance from the sun as about 4 to 9, the gravity of Saturn to Jupiter will be to the gravity of Saturn to the sun as 81 to $16 \times 1,067$, or about 1 to 211. And hence arises a perturbation of the orbit of Saturn at each conjunction of this planet with Jupiter that is perceptible enough that astronomers have difficulty with it. In accordance with the planet's changing position in these conjunctions, its eccentricity is now increased and now decreased, its aphelion is moved now forward and now back, and its mean motion is in turn accelerated and retarded. Nevertheless, nearly all the error in its motion around the sun arising from a force of this size (except for that in the mean motion) can be avoided by setting the lower focus of its orbit at the common center of gravity of Jupiter and the sun (by Book I Proposition 67); and therefore, where it is greatest, it hardly exceeds two minutes. And the greatest error in the mean motion hardly exceeds two minutes in a year. Further, in the conjunction of Jupiter and Saturn, the accelerative gravities of the sun towards Saturn, of Jupiter towards Saturn, and of Jupiter towards the sun, are approximately as $16, 81$, and $(16 \times 81 \times 3021)/25$, or $156,609$, and accordingly, the difference of the gravities of the sun towards Saturn and of Jupiter towards Saturn is to the gravity of Jupiter towards the sun as 65 to $156,609$, or 1 to $2,409$. And to this difference is proportional the greatest efficacy of

Saturn in perturbing Jupiter's motion, and consequently the perturbation of Jupiter's orbit is far less than that of Saturn's. The perturbations of the remaining orbits are much less again, except that the earth's orbit is perceptibly perturbed by the moon. The common center of gravity of the earth and the moon traverses an ellipse about the sun placed at the focus, and by a radius drawn to the sun describes areas proportional to the time on the same [ellipse]; the earth, however, revolves about this common center with a monthly motion.

Commentary on Book III: Propositions 9–42

Propositions 8 to 13, which you have just read, require little comment. Their results are striking and clear cut. After "weighing" the Sun, Earth, Jupiter, and Saturn in **Proposition 8**, Newton goes on in **Proposition 10** to provide assurance that the solar system is stable, that it will persist far into the future. In one sense, Propositions 11 to 13 refute the Copernican and Keplerian theories! **Proposition 12**, for example, shows that the Copernican claim that the Sun is the center of our system is not precise; what is at the center and stationary is the center of gravity of the entire solar system. **Proposition 13** goes beyond this to show that the Earth, for example, does not move in an ellipse; rather the *center* of the Earth-Moon system follows this path, except for the perturbations due to attraction from other planets. This section of the *Principia* illustrates a role that broad theories sometimes perform; they not only explain laws and phenomena, but also correct them, permitting a level of precision frequently beyond that which can be attained observationally.

The *Principia* by no means ends with Proposition 13 of Book III; it continues on to establish 29 more propositions for a total of 42 propositions. Some of these are of great importance; for example, in **Proposition 19**, Newton shows that the Earth is not a sphere, but is instead an oblate spheroid (like a tangerine) bulging slightly at the equator. A pendulum at the north pole must be lengthened slightly to swing with the same period it would have at the equator. **Propositions 24 through 38** deal with the motions of the Moon and also explain in this context the terrestrial tides. **Proposition 39** explains the precession of the equinoctial points and is followed by a lengthy treatment of comets, which Newton shows move in conic sections. The *Principia* of 1687 ended at this point, but to the 1713 second edition, Newton added a **General Scholium**, in which he gave expression to some of his deepest and most influential thoughts in regard to philosophy, religion, and science. The next (and final) selection from Newton's *Principia* consists of this General Scholium, which is widely regarded as one of the most interesting sections of Newton's masterpiece.

General Scholium[62]

The hypothesis of vortices is pressed by many difficulties. In order that any individual planet describe areas proportional to the time by a radius drawn to the sun, the periodic times of the parts of the vortex ought to have been in the duplicate ratio of the distances from the sun. In order that the periodic times of the planets be in the sesquiplicate ratio of the distances from the sun, the periodic times of the parts of the vortex ought to have been in the sesquiplicate ratio of the distances. In order that the smaller vortices about Saturn, Jupiter, and other planets be preserved in their circulations and float undisturbed in the sun's vortex, the periodic times of the parts of the solar vortex ought to have been equal. The rotations of the sun and the planets about their axes, which ought to have been consistent with the motions of the vortices, are in disagreement with all these ratios. The motions of the comets are in the highest degree regular, and observe the same laws as the motions of the planets, and cannot be explained by vortices. Comets are carried in highly eccentric motions into all parts of the heavens, which cannot happen unless vortices be removed.

Projectiles, in our air, feel only the resistance of our air. When the air is removed, as happens in Boyle's vacuum, the resistance stops, inasmuch as a slender feather and solid gold fall with equal velocity in that vacuum. And the account of the celestial spaces, which are above the sphere of the earth's exhalations, is the same. All bodies ought to move with complete freedom in these spaces, and for that reason the planets and comets revolve perpetually in orbits given in shape and position, following the laws set forth above. Though they will indeed carry on in their orbits by the laws of gravity, they nevertheless could by no means have attained the regular position of the orbits through these same laws.

The six principal planets revolve about the sun in circles concentric upon the sun, in the same direction of motion, approximately in the same plane. Ten moons revolve about the earth, Jupiter, and Saturn, in concentric circles, in the same direction of motion, very nearly in the planes of the orbits of the planets. And all these regular motions do not have their origin from mechanical causes, inasmuch as comets are carried freely in highly eccentric orbits, and to all parts of the heavens. In this kind of motion, the comets pass through the orbits of the planets with greatest ease and swiftness, and at their aphelia, where they

---

[62]Densmore, *Central Argument* (footnote 19), pp. 485–489.

are slower and delay for a longer time, they are at the greatest distance from each other, so that they pull each other least. This most elegant arrangement of the sun, planets, and comets could not have arisen but by the plan and rule of an intelligent and powerful being. And if the fixed stars be centers of similar systems, all these, constructed by a similar plan, will be under the rule of One, especially because the light of the fixed stars is of the same nature as the light of the sun, and all the systems send light into all mutually. And so that the systems of the fixed stars should not fall into each other mutually, he will have placed this same immense distance among them.

He governs everything, not as the soul of the world, but as lord of all things. And because of his dominion, he is usually called "Lord God Παντοκρατωρ."[63] For "God" is a relative word, and is related back to servants, and "deity" is the absolute rule of God, not over his own body, as those believe for whom God is the world soul, but over servants. God most high is a being eternal, infinite, absolutely perfect; but a being without dominion, however perfect, is not the Lord God. For we say, "my God," "your God," "God of the Israelites," "God of gods," but we do not say, "my eternal," "your eternal," "eternal of the Israelites," "eternal of gods;" we do not say, "my infinite," or "my perfect." These names have no relation to servants. The word "God" everywhere signifies[64] the Lord; but not every lord is God. The absolute rule of a spiritual being constitutes God: true [absolute rule constitutes] the true [God]; the highest [absolute rule constitutes] the highest [God]; sham [absolute rule constitutes] a sham [God]. And from true absolute rule it follows that the true God is living, intelligent, and powerful; from the remaining perfections, that he is the highest, or in the highest degree perfect. He is eternal and infinite, omnipotent and omniscient; that is, he endures from eternity to eternity, and is present from infinity to infinity. He reigns over everything, and knows everything that happens or can happen. He is not eternity and infinity, but eternal and infinite; he is not duration and space, but he endures and is present. He endures always, and is present everywhere, and by existing always and everywhere, he has established duration and space. Since any single particle whatever of space is always, and any single indivisible moment whatever of duration is

---

[63][Newton's marginal note:] That is, "Universal Emperor."

[64][Newton's marginal note:] Our Pocock derives the word "deus" from the Arabic word "du" (and in the oblique case, "di"), which signifies the Lord. And in this sense, princes are called "dii", Psalm 84:6 and John 10:45. And Moses is called "deus" of his brother Aaron, and "deus" of king Pharaoh (Exodus 4:16 and 7:1). And in the same sense the souls of dead princes used to be called "dii" by the people, but falsely, on account of the want of dominion.

everywhere, surely the maker and lord of all things will not be *never, nowhere.* Every sentient soul is the same indivisible person at different times and in the different organs of perception and motion. Successive parts are given in duration, coexisting parts in space, neither [is given] in the human person or in his thinking principle: much less so in the thinking substance of God. Every person, *qua* sentient thing, is one and the same person throughout life in each and every organ of perception. God is one and the same God always and everywhere. He is omnipresent not in *power* alone, but also in *substance.* For power cannot subsist without substance. In him all things are contained[65] and moved, but without mutual effects [*passio*]. God is not affected by the motions of bodies, and these do not experience any resistance from God's omnipresence. It is universally acknowledged that the highest God exists necessarily, and by the same necessity he is *always* and *everywhere.* Hence also, the whole is entirely similar to himself, all eye, all ear, all brain, all force of perceiving, understanding, and acting, but in a manner by no means human, in a manner by no means corporeal, in a manner entirely unknown to us. Just as a blind man has no idea of colors, so we have no idea of the ways in which God most wise perceives and understands everything. He is entirely void of all body and corporeal form, and therefore cannot be seen, nor heard, nor touched; nor ought he to be worshipped under the image of any corporeal object. We have ideas of its[66] attributes, but we do not have the least knowledge of what the substance of any object is. We see only the shapes and colors of bodies, we hear only sounds, we touch only the external surfaces, we smell only odors, and taste flavors: we have no cognition of the inmost substances by any sense or act of reflection, and much less do we have an idea of the substance of God. We have cognition of him only through his properties and attributes, and

---

[65][Newton's marginal note:] This was the opinion of the ancients, such as Pythagoras (in Cicero, *On the nature of the gods* Book 1, Thales, Anaxagoras, Virgil (*Georgics* 4:220, and *Aeneid* 6:721), Philo (*Allegories,* beginning of Book 1), Aratus (*Phenomena,* at the beginning). So also the sacred writers, such as Paul (*Acts* 17:27–28), John 14:2, Moses (*Deuteronomy* 4:39 and 10:14), David (*Psalm* 139:7, 8, 9), Solomon (1 *Kings* 8:27), *Job* 22:12, 13, 14, Jeremiah 23:23–24. Moreover, idolaters used to make out that the sun, the moon, and the stars, people's souls, and other parts of the world, are parts of the highest god, and are therefore to be worshipped, but falsely.

[66]The word Newton uses here, "*eius*", could be masculine or feminine or neuter; hence, it is not possible to tell whether Newton means God's attributes or those of bodies. In the context of the preceding sentence, this word would refer to God, but in the context of what follows, it would refer to bodies. Although the latter translation has been chosen, as being more consistent with the argument, the former is not impossible, and was adopted by Motte. [translator's note]

through the wisest and best structures and final causes[67] of things, and marvel because of [his] perfections, and further, we revere and worship [him] because of his dominion. For we worship as servants, and God without dominion, providence, and final causes is nothing different from fate and nature. From blind metaphysical necessity, which is absolutely the same always and everywhere, no variation of things arises. The whole diversity of created things according to places and times could only have arisen from the ideas and will of a being existing necessarily. Moreover, God is said by way of allegory to see, to speak, to laugh, to love, to hate, to desire, to give, to receive, to rejoice, to become angry, to fight, to devise, to establish, to build. For every account of God is taken from human things through a certain likeness, not indeed perfect, but of a certain sort. And this much concerning God, to discourse of whom, at least from the phenomena, is the business of natural philosophy.

Hitherto I have set forth the phenomena of the heavens and of our sea through the force of gravity, but I have not yet assigned the cause of gravity. This force does indeed arise from some cause, which penetrates all the way to the centers of the sun and the planets, with no diminution of power, and which acts, not according to the quantity of the *surfaces* of the particles upon which it acts (as mechanical causes are wont to do), but according to the quantity of solid matter, and which acts at immense distances, extended everywhere, always decreasing in the duplicate ratio of the distances. Gravity towards the sun is compounded of the gravities towards the individual particles of the sun, and in receding from the sun decreases precisely in the duplicate ratio of the distances all the way to the orbit of Saturn, as is manifest from the planets' aphelia being at rest, and all the way to the aphelia of the comets, provided that those aphelia are at rest. The reason for these properties of gravity, however, I have not yet been able to deduce from the phenomena, and I do not contrive hypotheses. For whatever is not deduced from the phenomena is to be called a *hypothesis*, and hypotheses, whether metaphysical or physical, whether of occult qualities or mechanical ones, have no place in experimental philosophy. In this philosophy, propositions are deduced from the phenomena, and are rendered general by induction. Thus the impenetrability, mobility, and impetus of bodies, and the laws of motions and of gravity, came to be known. And it is enough that gravity really exists, and acts according to the laws set forth by us, and is sufficient

---

[67]"Final cause" is an Aristotelian term that denotes the end for the sake of which something happens. See Aristotle, *Physics*, II.3, 194b 32. [translator's note]

[to explain] all the motions of the heavenly bodies and of our sea.

It would now be appropriate to add some remarks about a certain extremely subtle spirit pervading gross bodies and lying hidden in them, by whose force and actions the particles of bodies attract each other mutually at least distances, and stick together when brought into contact, and electrical bodies act at greater distances, both repelling and attracting neighboring corpuscles, and light is emitted, reflected, refracted, inflected, and heats bodies, and all perception is aroused, and the members of animals are moved by the will, that is, by vibrations of this spirit propagated through the solid filaments of the nerves from the external organs of perception to the cerebrum and from the cerebrum to the muscles. But these cannot be set forth in a few words, nor is there at hand a sufficient body of experiments by which the laws of action of this spirit are required to be accurately determined and shown.

Commentary on Newton's General Scholium

That Newton was a religious person intensely concerned about theological issues is suggested by the middle two of the six paragraphs in his General Scholium. This point is further documented by the fact that among his manuscripts are some 1,300,000 words on theological issues. By birth an Anglican, Newton nonetheless acquired anti-Trinitarian and Arian leanings that he did not wish to make public. This feature of his thought as well as his passion for alchemical investigations (650,000 words in his manuscripts) have become most widely known and studied only in the twentieth century as his unpublished writings have been analyzed. An important issue to raise is whether Newton in his General Scholium portrays God as remote and impersonal. One feature that is definitely clear is Newton's strong conviction that the study of the laws of nature should be recognized as in some sense a religious activity that reveals, among other points, the majesty of the Almighty.

In the next to last paragraph of the General Scholium, Newton points out that he does not consider gravitational attraction as an ultimate mode of explanation. As he states: "The reason for these properties of gravity, however, I have not been able to deduce from the phenomena...." In this regard, he agreed with such quasi-Cartesians as Christiaan Huygens and Gottfried Wilhelm Leibniz, who argued that gravity itself requires an explanation. Newton could not, however, agree with explanations of gravity that they proposed. Nor was Newton sympathetic with the tendency of some Cartesians to disparage the *Principia* on the grounds that gravitation had been left unexplained. To such persons Newton directed his remark: "And it is enough that gravity really exists, and acts according to the laws set forth by us, and is sufficient [to explain] all the motions of the heavenly bodies and of our sea."

Newton's next to last paragraph contains one of his most famous and most misunderstood statements. In his Latin text, Newton had written "hypotheses non fingo," which Andrew Motte, the first English translator of the *Principia*, erroneously translated as "I frame no hypotheses." Solid evidence indicates that Newton's Latin should be translated: "I do not contrive hypotheses" or as is sometimes done, "I do not feign hypotheses." By this he meant to express his opposition, not to formulating hypotheses, but to "contriving" or "feigning" *artificial* hypotheses that, although they account for the phenomena, are known not to be true. This is but another evidence that Newton, like his great predecessors Copernicus, Kepler, and Galileo, had strong leanings toward a realist position.

The Structure of Newton's *Principia* and the Hypothetico-Deductive Method

Introduction

A number of philosophers have claimed that most or all scientific arguments take the form of the "hypothetico-deductive method." This section consists of two parts: (1) an examination of the nature and history of this method, and (2) the suggestion that an awareness of this method can provide a illuminating interpretation of the structure of Newton's *Principia*.

Preliminary Logical Discussion

Examine each of the following syllogisms to determine its validity:

1) If A, then B.	2) If A, then B.	3) If A, then B.
But A.	But not B.	But B.
Therefore B.	Therefore not A.	Therefore A.

The first syllogism is clearly valid. For example, if we know that if it rains, the grass is wet, then we can validly conclude from the information that it has rained, that the grass is wet. The second is also valid. If we know that if it rains, the grass is wet, then we can validly conclude from the fact that the grass is not wet, that it has not rained. Syllogism #3 is not, however, logically valid. Assuming that we know that if it rains, the grass is wet, we cannot validly conclude from the fact that the grass is wet, that it has rained. A sprinkler might have produced that result.

Can Either Deduction or Induction Be Claimed as the Sole Scientific Method?

One problem with the claim that the method of science is induction is that scientific explanations often refer to various hypothetical entities, for example, atoms, strange forces, or epicycles that are not directly observable. Moreover, it has become increasingly clear from historical studies that detailed empirical knowledge does not always produce correct theories and that frequently the appearance of new theories has not been a direct result of the discovery of new empirical information. On the other hand, it is surely problematic to argue that deduction is the sole method of science. Such a claim seems to entail the necessity of specifying a source establishing the certainty. or at least the probability, of the premises used in the deductions. Moreover, this view ill accords with the fact that observational and experimental data have a central and crucial role in science.

The Hypothetico-Deductive Method

Because it is difficult to maintain that induction alone or deduction alone is the method of science, methodologists of science have at times favored a method that in a sense combines both. This is the hypothetico-deductive method (hereafter HD method). Proponents of the HD method claim that scientists begin with hypotheses. Such proponents usually leave aside the question of the source of these hypotheses, suggesting that an answer to this question is not necessary for specifying the method used by scientists for verifying their claims. Moreover, it is typically asserted that the process of discovery does not admit of logical analysis, that there cannot be a logic of discovery, but at most only a logic of verification and falsification. Once a scientist attains an hypothesis, he or she deduces conclusions from it, possibly using in this process other relevant information, e.g., initial conditions. These deductions characteristically take the form of predicting phenomena. Looked at formally, the process begins with H implies $p_1, p_2, p_3, \ldots, p_n$, where H is an hypothesis and $p_1, p_2, p_3, \ldots, p_n$ are various phenomena. The scientist then investigates whether these phenomena occur. If they do in fact occur, then one concludes that there is good evidence for the hypothesis. On the other hand, if the scientist finds "not p," that is, if one or more of the deduced phenomena does not occur, then the conclusion is that the hypothesis is false. In short, the two forms of argument are:

$1')\ H \Rightarrow p_1, p_2, p_3, \ldots, p_n.$
But $p_1, p_2, p_3, \ldots, p_n.$
Therefore H.

$2')\ H \Rightarrow p_1, p_2, p_3, \ldots, p_n.$
But not $p_x.$
Therefore not H.

As an illustration of this, consider the following analysis. If the Darwinian theory of evolution by natural selection is true, then one should find a fossil record of a certain sort, vestigial organs, various similarities among animals, certain patterns of ecological distribution, etc. If these deductively derived phenomena are found to occur, then the theory deserves credibility, whereas if one finds that one or more does not occur, then the theory is falsified.

Huygens and the HD Method

Perhaps the earliest well formulated statement of the HD method occurs in Christiaan Huygens's *Treatise on Light* (1690). In that work, Huygens put forward his theory that light consists in pulses in the ether. In the preface to his book, Huygens discussed the method of argumentation employed in his book to support this theory; in particular, Huygens stated:

> There will be seen in [this book] demonstrations of those kinds which do not produce as great a certitude as those of Geometry, and which even differ much therefrom, since whereas the Geometers prove their Propositions by fixed and incontestable Principles, here the Principles are verified by the conclusions to be drawn from them; the nature of these things not allowing of this being done otherwise. It is always possible to attain thereby to a degree of probability which very often is scarcely less than complete proof. To wit, when things which have been demonstrated by the Principles that have been assumed correspond perfectly to the phenomena which experiment has brought under observation; especially when there are a great number of them, and further, principally, when one can imagine and foresee new phenomena which ought to follow from the hypotheses which one employs, and when one finds that therein the fact corresponds to our prevision. But if all these proofs of probability are met with in that which I propose to discuss, as it seems to me they are, this ought to be very strong confirmation of the success of my inquiry; and it must be ill if the facts are not pretty much as I represent them.[68]

---

[68] Christiaan Huygens, *Treatise on Light*, trans. by Silvanus P. Thompson (Dover Publications, n.d.), pp. vi–vii.

Advantages and Problems of the HD Method

One major advantage of the claim that science employs the HD method is that this claim accounts for the fact that both empirical information and deductive inferences occur repeatedly in science. According to the HD method, the deductions characteristically take the form of deducing empirical information from the hypotheses. Empirical information also plays a vital role: it is used to check the hypotheses; in particular, comparisons between observed phenomena and deductions from the hypotheses test the hypotheses. There is, however, an apparent problem: the form of the HD method of inference is that represented above as case #3: "If A, then B. But B. Therefore A." The problem is that this syllogistic form is invalid. The source of its invalidity is that there can exist another hypothesis, call it A', that also entails B as a consequence or, to put it differently, another hypothesis may deductively yield the same conclusions or phenomena. What responses to this objection can be made? One response is that provided by Huygens: "the nature of these things [does] not allow. . . of this being done otherwise." He is saying, in other words, that because scientists cannot, like geometers, attain "fixed and incontestable Principles," they must have recourse to this method and be content with the probable conclusions that can be drawn by it. (Note that when the HD method is used in form #2, that is, in falsifications, it is valid). Moreover, Huygens places a number of conditions on this method: a "great number" of phenomena must be found and must be deducible from the hypothesis and the hypothesis should predict some new phenomena.

Conclusion

Various philosophers have championed the claim that the main method of science is the HD method. For example, during the nineteenth century, it was advocated by William Stanley Jevons, who referred to it as the method of "inverse deduction," and by Charles Sanders Peirce, who sometimes called it "abduction" and at other times "retroduction." A full blown argument for it would entail specifying more precisely the criteria to be used in assessing the probability of hypotheses and possibly some modification of the claim that a single counterinstance necessitates abandoning the hypothesis. It is also important to recognize that in many cases where this form of argument is used, the goal is to argue not for a single hypothesis but rather for a cluster of related hypotheses. It may also be the case that the phenomena are justified by means of a string of deductions, not simply by one deduction.

Newton and the Hypothetico-Deductive Method

What Is the Methodological Structure of Newton's *Principia*?

It is clear that both deductive and inductive arguments appear in the *Principia*. Nonetheless, it is important to ask: what is the overall structure of the *Principia*? Quite discordant answers to this question have been given by prominent scholars. For example, the eminent physicist and historian and philosopher of science Pierre Duhem (1861–1916) has discussed what he called the "Newtonian method," which he describes as an ideal to which many physicists aspire. Duhem asserts that according to this method, the physicist wishing to formulate a physical theory should follow this procedure:

> The hypotheses from which [the physical theory] starts and develops its conclusions would...be tested one by one; none would have to be accepted until it presented all the certainty that experimental method can confer on an abstract and general proposition; that is to say, each would necessarily be either a law drawn from observation by the sole use of these two intellectual operations called induction and generalization, or else a corollary mathematically deduced from such laws. A theory based on such hypotheses would then not present anything arbitrary or doubtful....[69]

Duhem proceeds to explain why he calls this the Newtonian method: "It was this sort of physical theory that Newton had in mind when, in the General Scholium which crowns his *Principia*, he rejected so vigorously as outside of natural philosophy any hypothesis that induction did not extract from experiment; when he asserted that in a sound physics every proposition should be drawn from phenomena and generalized by induction."[70] The passage from Newton's General Scholium that Duhem had in mind was no doubt the following: "In this [my experimental] philosophy, propositions are deduced from the phenomena, and are rendered general by induction. Thus...the laws of motion and of gravity came to be known."[71] Duhem went on to document that the eminent electrical theorist André Marie Ampère (1775–1836), sometimes called the "Newton of electricity," also ascribed this method to Newton and attempted to model his publications on electrical theory on this method.[72]

---

[69] Pierre Duhem, *The Aim and Structure of Physical Theory*, trans. by Philip P. Wiener (Princeton University Press, 1954), p. 190.

[70] Duhem, *Physical Theory*, pp. 190–191.

[71] Densmore, *Central Argument* (footnote 19), p. 488.

[72] Duhem, *Physical Theory*, pp. 195–196.

What is striking is that Duhem then proceeds to argue that in effect the Newtonian method is a myth, that in fact physical theory cannot be constructed in such a manner. Moreover, he argues that Newton himself had not followed this method in his *Principia*. He does this by challenging the claim that Newton derived his theory of universal gravitation from Kepler's laws. He sums up his case by claiming: *"The principle of universal gravity, very far from being derivable by generalization and induction from observational laws of Kepler, formally contradicts these laws. If Newton's theory is correct, Kepler's laws are necessarily false."*[73] In particular, Duhem notes that Newton himself proves that planets do not move precisely in ellipses because of the perturbations caused by the attraction of other planets. Thus it becomes problematic how Newton could derive his theory partly on the basis of Kepler's first law if that law is itself only approximate. Duhem's claim in this regard was endorsed by Karl Popper, one the most important authorities on scientific method of the twentieth century, when Popper wrote:

> It is well known that Newton's dynamics achieved a unification of Galileo's terrestrial and Kepler's celestial physics. It is often said that Newton's dynamics can be induced from Galileo's and Kepler's laws, and it is even been asserted that it can strictly be deduced from them. But this is not so; from a logical point of view, Newton's theory, strictly speaking, contradicts both Galileo's and Kepler's (although these latter theories can of course be obtained as approximations, once we have Newton's theory to work with). For this reason it is impossible to derive Newton's theory from either Galileo's or Kepler's or both, whether by deduction or induction. For neither a deductive nor an inductive inference can ever proceed from consistent premises to a conclusion that formally contradicts the premises from which we started.[74]

The chief point that I wish to draw from this discussion of Duhem and Popper is this: whatever method Newton followed in his *Principia* it was not the method that Duhem and Ampère described as the "Newtonian method." Or to put the point differently, Newton's famous statement in his General Scholium should be seen as a misstatement by Newton of how he attained such remarkable successes.

Newton's View of the HD Method

Various passages in Newton's writings make clear that he was aware of the HD method. These passages, however, also show that he opposed it and

---

[73] Duhem, *Physical Theory*, p. 193.
[74] Karl Popper, *Objective Knowledge: An Evolutionary Approach* (Clarendon, 1972), p. 198.

claimed that he had found a method of circumventing its inherent difficulties. For example, Newton on 6 July 1672 wrote Henry Oldenberg:

> I cannot think it effectual for determining truth, to examine the several ways by which the phaenomena may be explained, unless there can be perfect enumeration of all those ways. You know the proper method for inquiring after the properties of things is to deduce them from experiments. And I told you that the theory which I propounded was evinced by me, not by inferring, *it is thus, because it is not otherwise*; that is, not by deducing it only from a confutation of contrary suppositions, but by deriving it from experiments concluding positively & directly.[75]

This passage and others show that Newton (always aware that his physical ideas were in competition with Cartesian modes of explanation) was very anxious that his presentations not be seen as HD in form, and thus be dismissed as just another theory. It is, however, quite another question whether Newton successfully avoided using the HD method. In fact, the next section contains the suggestion that the HD method can provide an interpretation of the structure and the overall significance of Newton's *Principia*.[76]

The Methodology of Newton's *Principia*

As noted repeatedly in the earlier sections of this chapter, one of Newton's chief concerns was to work out the relations between the three Keplerian laws of motion and his mechanics. Moreover, Newton was intent on deriving from his mechanics the results that Galileo had achieved. A good example of this is that Newton, by showing that bodies acting under an inverse square field of force move in either an ellipse, parabola, or hyperbola,

---

[75] As quoted in Ralph M. Blake, "Isaac Newton and the Hypothetico-Deductive Method" in Ralph M. Blake, Curt J. Ducasse, and Edward H. Madden (eds.), *Theories of Scientific Method: The Renaissance through the Nineteenth Century* (University of Washington Press, 1960), p. 127. See *Isaaci Newtoni Opera quae exstant omnia,* ed. by Samuel Horsley, vol. 4 (J. Nichols, 1782), p. 320.

[76] Some authorities reject the idea that Newton used the HD method. One such Newtonian expert is William Harper, who nonetheless admits early in his analysis that "Duhem, Popper, and Lakatos insist. . . that only hypothetico-deductive construal of Newton's evidence for universal gravitation makes sense." (See William Harper, "Newton's Argument for Universal Gravitation" in I. Bernard Cohen and George Smith (eds.), *The Cambridge Companion to Newton* (Cambridge University Press, 2002), pp. 174–201:174.) Not only did these three eminent philosophers claim that Newton used this method, but also the *Encyclopædia Britannica* in its online article "Hypothetico-Deductive Method" attributes the creation of the method to Newton! My concern at this point is not, however, to argue for the method but only to use it to illuminate the structure of Newton's *Principia*.

was accounting for the truth of Galileo's claim that projectiles on the Earth follow a parabolic path. In addition, by combining Newton's second law of motion with his law of gravitation, it becomes evident why Galileo was correct in claiming that the weight of a body in free fall does not affect its rate of descent. It was also a very impressive feature of Newton's system that he could offer an explanation of the precession of the equinoxes and of tidal phenomena. The situation can be roughly diagrammed in the following manner. The box on the left contains the main theoretical elements that Newton presents in Book I and from which he deduces (also in Book I) a large number of very general principles and theorems. The elements in the box on the right are laws or other empirical results that Newton deals with in Book III.

Definitions **Newton's 1st Law of Motion** **Newton's 2nd Law of Motion** **Newton's 3rd Law of Motion**	**Kepler's 2nd Law of Motion** **Galileo's Results** **Parabolic path of projectiles** **Precession of the equinoxes** **Tidal phenomena** **Cometary orbits** **etc.**

This diagram needs to be qualified in at least two ways. First, this diagram does not do justice to the fact that Newton developed long deductive chains, especially in Book I. Second, this list should not be assumed to be exhaustive; for example, Newton's views on absolute space and time were part of the fundamental structure of his system. Moreover, Newton's experiments recounted in Book III supporting the equality of inertial and gravitational mass were crucial for what followed. Similarly, Newton relied so heavily on his "Rules of Philosophizing" that they are candidates for inclusion in the box on the left.

Although this diagram on one level correctly represents the structure of the argument that Newton offers his readers, this interpretation can easily lead to confusion about the structure of Newton's overall argument. One problematic aspect of this representation of Newton's method is that it portrays Newton as using his laws of motion to prove results that were far more certain than the laws from which he was proving them. As must be clear, by 1687, Kepler's second law was widely known and had been supported by substantial empirical evidence. The same is true for the central results of Galilean mechanics. These laws and results were far more certain and much more widely accepted than Newton's second and third laws of motion. Even Newton's law of inertia was in some ways problematic, two of the main reasons being that one could never remove all forces from a body, nor even if one somehow managed to remove these forces, could one observe the body's motion eternally. How is this puzzle—that Newton seems to be trying to prove the better known by the less well known—to be resolved? Moreover, it is puzzling that Newton does not offer much empirical evidence for his laws of motion.

A solution to these puzzles comes from a consideration of the HD method, which says that hypotheses are tested by their success in providing predictions and deductions of empirical results. Thus, although on one level, the argument in the *Principia* flows from Book I to Book III, in reality the argument flows in the opposite direction. Newton's laws are warranted by the fact that from them, one can generate a large number of known results. Thus, in effect, the phenomena set out in Book III of the *Principia* confirm claims that had been set out in Book I. This is represented in the next diagram.

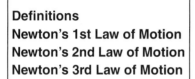

Definitions
Newton's 1st Law of Motion
Newton's 2nd Law of Motion
Newton's 3rd Law of Motion

Kepler's 2nd Law of Motion
Galileo's Results
Parabolic path of projectiles
Precession of the equinoxes

Tidal phenomena
Cometary orbits
etc.

In other words, the suggestion is that it is appropriate to view the *Principia* as having a HD structure. The hypothetical part consists of Newton's laws and definitions; the phenomenal part consists of such empirical matters as Kepler's second law, Galileo's results, as well as precessional, tidal, and cometary phenomena. The warrant for accepting Newton's laws is that from them these phenomena can be deduced. Finding that these phenomena occur shows that Newton's laws are acceptable.

An important question that arises in this context concerns the relation of Kepler's laws to Newtonian mechanics. The fact that answers given to this question by various authors differ somewhat suggests the complexity of the question. This much, however, can be said. It is true that from Newton's three laws of motion and the law of universal gravitation, one can mathematically deduce Kepler's three laws of motion (leaving aside considerations of perturbations). It is, however, untrue that this is what Newton in fact did in his *Principia*. It is correct that from Newton's laws of motion, he did deduce Kepler's second law. On the other hand, Newton used Kepler's first and third laws in his derivation of his inverse square law of universal gravitation. Moreover, he proceeded to use the law of gravitation to prove, and correct, Kepler's first law. It is also true that Kepler's first and third laws form a sort of triad with the inverse square law: from any one of the three when combined with Newton's laws of the motion, the other two members of the triad can be proven. It also seems probable that historically the realization that from Newton's three laws of motion and the law of universal gravitation one can deduce Kepler's laws generated support for the Newtonian system. It should be noted, however, that my argument that is possible to see a HD structure in the *Principia* does not rest on this possibility. These comments should also make clear why Kepler's first and third laws do not appear in the box on the right in the accompanying diagram.

The philosopher of science Stephen Toulmin has presented the following interesting analysis as a basis for seeing Newton's *Principia* as exhibiting a HD structure. Part of Toulmin's approach consists in comparing and contrasting Newton's methodology with that advocated by Descartes.

> Granted that the theory of motion and gravitation of Newton's *Principia* did indeed conform to Descartes' recipe—adding further dynamical axioms, definitions, and postulates to those of Euclid's geometry— Newton nonetheless made no pretense of proving, in advance of empirical evidence, that these additional assumptions were uniquely self-evident and valid. Instead, he treated them as working assumptions to be accepted hypothetically for just so long as their consequences threw light, in exact detail, on hitherto-unexplained phenomena. Inevitably, the epistemic claims to be made on behalf of such explanations fell short of Descartes's full "deductivist" ambitions. Newton knew of no phenomena, for instance, that evinced the mechanisms of gravitational attraction and saw no point in 'feigning hypotheses' about them.
>
> In this way, Newton devised in practice—almost inadvertently—what philosophers of science have since labelled the hypothetico-deductive method, in which...the proper form of a theory is seen as a mathematical system in which particular empirical phenomena are explained by relating them back deductively to a small number of general principles and definitions. The method, however, abandons the Cartesian claim that those principles and definitions can themselves be established, finally and conclusively, before inquiring what light their consequences throw on actual scientific problems and phenomena.[77]

Another mystery about the Newtonian system that an analysis in terms of the HD method can illuminate relates to the Copernican theory. At the beginning of the seventeenth century, the majority of scientists doubted the truth of the Copernican system, one reason being the failure to detect stellar parallax. By the early eighteenth century, although no conclusive evidence of parallax had been found, nearly all scientists accepted Copernicanism. What had brought about this change? It was not because astronomers had

---

[77]Stephen Toulmin, "Science, Philosophy of," *Encyclopædia Britannica,* vol. 16, 15th ed. (Encyclopædia Britannica, Inc., 1983), pp. 375–393: 378–379.

attained some direct proof of the Copernican system. Parallax, for example, was first measured only in 1838. A major factor in the increased acceptance of the Copernican system was that it had become an integral part of the systems created by Kepler, Galileo, Descartes, and Newton. Although it is true that the Galilean, Cartesian, and Newtonian advances in mechanics increased the acceptability of the Copernican system, the Copernican system in many ways provided insights essential to the Galilean, Cartesian, and Newtonian developments in mechanics. For example, whereas such insights as the law of inertia provided Copernicans with an explanation of why objects on a moving Earth are not left behind, the Copernican theory provided suggestions as to the causes of tidal phenomena. This supports the idea that the new mechanics and Copernican astronomy evolved in tandem. It was not that one proved the other, freed it from all difficulties, or suggested new approaches, but rather that each assisted the other in those ways. In this context, it is interesting to consider the fact that Halley shortly before the *Principia's* publication had informed the Royal Society that Newton's treatise "gives a mathematical demonstration of the Copernican hypothesis."[78] Because the Newtonian system at that time (1686) had not yet even been published whereas the Copernican theory had been winning adherents for 143 years, it might seem strange that an author would try to prove the much older and better tested system from the new system. The HD method clarifies this issue: it suggests that a powerful test of a theory is its ability to generate true results. And thus Halley was in effect suggesting that one important piece of evidence for Newton's new system was its ability to deduce the Copernicanism system.

This overall analysis simultaneously suggests a way around a claim made around 1900 by Henri Poincaré, who asserted that the law of inertia is incapable of empirical proof. Pierre Duhem countered that claim by asserting in his *Aim and Structure of Physical Theory* (1906) that physical theories, such as the law of inertia, are never subjected to test in isolation; rather theories always go to test in a group.[79] If one or more of the phenomena predicted by the group of theories does not occur, what this shows is that one or more of the theories in the group is defective. Unfortunately, the negative result does not tell us which theory is faulty. A good example of this is the Copernican theory. It cannot be tested in isolation. Think of the problem of parallax. Copernicus dealt with that challenge by noting that the failure to find parallax could be explained by the stars being so distant that no parallax is observable. To put it differently, Copernicus realized that his system was actually going to the parallax test in conjunction with an assumption

---

[78] As quoted in Westfall, *Never at Rest,* pp. 444–445
[79] Duhem, *Physical Theory,* pp. 212–216.

about the relative nearness of the stars. He saved his system by proposing that this additional assumption might have to be discarded.

The goal of this section has not been to argue that the main method of science is the HD method. My chief concerns in presenting this analysis have been to suggest one plausible view of the structure of the *Principia* as a whole and also to point out the problematic character of a widespread misrepresentation of the method Newton employed. This misrepresentation consists in the assumption that Newton's *Principia* should be read as a book in which each law or proposition is inductively justified at the time when it is presented. My argument in this regard is that Newton's *Principia* should not and cannot be read in an inductivist manner. The reading here proposed leaves open the possibility of various other interpretations of the *Principia,* one of which is the HD reading.

Isaac Newton's Correspondence in 1692–1693 with Rev. Richard Bentley

Introduction

Isaac Newton was in his mid-forties and largely unknown except for some of his optical writings when in 1687 he published his *Principia*. Thus it is not surprising that his masterpiece only gradually came to be seen as perhaps the most remarkable production of the seventeenth century. Few of Newton's contemporaries recognized the genius manifested in Newton's volume earlier than Rev. Richard Bentley (1662–1742), a Cambridge educated divine, who in 1691 was serving as chaplain to the Bishop of Worcester. Although Bentley was known above all as a classicist, in fact, one of the most eminent classicists of that age, he decided in 1691 to attempt reading the *Principia*. Seeking advice from the mathematician John Craige on what he would need to read before proceeding to the *Principia,* Bentley received a staggeringly long list of readings. Undaunted, he turned to Newton, who supplied a shorter list and suggested ways in which Bentley could get to the core of Newton's text by leaving various parts aside. Aided in this way, Bentley tackled the *Principia*.[80]

The learning Bentley thereby attained proved of great value when in 1692 Bentley was chosen to inaugurate a series of sermons funded by the famous chemist Robert Boyle, who had died in 1691. Boyle, who was very

---

[80]On Newton's relations to Bentley, see Perry Miller "Bentley and Newton" in I. Bernard Cohen (ed.), *Isaac Newton's Papers and Letters on Natural Philosophy* (Harvard University Press, 1958), pp. 271–278. This volume also contains Bentley's seventh and eighth Boyle lectures, those in which he drew on Newton's ideas.

concerned about the relations between science and religion, had endowed the series under the following arrangement:

> To settle an annual salary for some divine or preaching minister, who shall be enjoined to perform the offices following: 1. To preach eight Sermons in the year, for proving the Christian religion against notorious infidels, *viz.*, Atheists, Deists, Pagans, Jews, and Mahometans; not descending to any controversies that are among Christians themselves. . . .[81]

Bentley, titling his sermons *The Confutation of Atheism,* described their contents on his title page:

> The Folly and Unreasonableness of Atheism demonstrated from The Advantage and Pleasure of a Religious Life, The Faculties of Human Souls, The Structure of Animate Bodies, and The Origin and Frame of the World, etc.[82]

Bentley's chief target in his attack was Thomas Hobbes (1588–1679), who was himself a proponent of the materialism of such ancient atomists as Lucretius. Bentley, believing that Newton might find evidence in his system in support of belief in a divinity, wrote to Newton, raising various questions. Newton responded warmly with four letters. In these letters Newton revealed many of his most significant ideas regarding how his system could be related to religion. Bentley drew heavily on his reading of the *Principia* and also on these four letters in the concluding seventh and eighth sermons of his *Confutation of Atheism,* which was judged to be a great success.

Bentley kept in contact with Newton, especially after 1700, when Bentley was named Master of probably the most prominent Cambridge college, Trinity College, which was Newton's college. Bentley remained Master of Trinity until his death in 1742. Bentley in fact played an important role in persuading Newton to prepare the 1713 second edition of his *Principia,* for which Newton wrote his General Scholium.

In reading Newton's four letters to Bentley, you may wish to keep in mind the following questions.

1. Did Newton believe that the universe that he had worked out was supportive of or hostile to religion?

2. Newton's notion of gravitation might, it seems, have been seen as supporting materialistic, atomistic conceptions of the universe. For

---

[81] As quoted in H. S. Thayer (ed.), *Newton's Philosophy of Nature: Selections from His Writings* (Hafner, 1953), p. 188.

[82] As quoted in Thayer, *Newton's Philosophy,* p. 188.

example, some persons speculated that our solar system could have formed by the action of gravity on the material in the universe. What was Newton's position on this issue? What were the chief evidences he cited in support for his position?

3. If we suppose the universe to be finite and that matter is scattered throughout it and that all that matter is mutually attractive, we could easily conclude that the universe must eventually under gravitational attraction collapse upon itself. How does Newton deal with this problem?

4. What was Newton's personal view of the nature of gravity? Did he view it as, for example, innate in matter? Did he view it as an explanatory category or as something itself in need of explanation?

Published in 1756, Newton's very influential four letters to Richard Bentley are given below in their entirety.

Newton's Letters to Bentley[83]

LETTER I

Sir,

When I wrote my Treatise about our System, I had an eye upon such principles as might work with considering men for the belief of a Deity; and nothing can rejoice me more than to find it useful for that purpose. But if I have done the public any service this way, it is due to nothing but industry and patient thought.

As to your first Query, it seems to me, that if the matter of our sun and planets, and all the matter of the universe, were evenly scattered throughout all the heavens, and every particle had an innate gravity toward all the rest, and the whole space, throughout which this matter was scattered, was but finite, the matter on the outside of this space would by its gravity tend towards all the matter on the inside and by consequence fall down into the middle of the whole space, and there compose one great spherical mass. But if the matter was evenly disposed throughout an infinite space, it could never convene into one mass; but some of it would convene into one mass and some

---

[83] Isaac Newton's four letters to Richard Bentley taken from *Isaaci Newtoni Opera quae exstant omnia*, ed. by Samuel Horsley, vol. 4 (J. Nichols, 1782), pp. 429–442. Such matters as spelling and punctuation have been slightly modernized.

into another, so as to make an infinite number of great masses, scattered at great distances from one to another throughout all that infinite space. And thus might the sun and fixed stars be formed, supposing the matter were of lucid nature. But how the matter should divide itself into two sorts, and that part of it, which is fit to compose a shining body, should fall down into one mass and make a sun; and the rest, which is fit to compose an opaque body, should coalesce, not into one great body, like the shining matter, but into many little ones; Or if the sun at first were an opaque body like the planets or the planets lucid bodies like the sun, how he alone should be changed into a shining body whilst all they continue opaque, or all they be changed into opaque ones whilst he remains unchanged, I do not think explicable by mere natural causes, but am forced to ascribe it to the counsel and contrivance of a voluntary Agent.

The same power, whether natural or supernatural, which placed the sun in the center of the six primary planets, placed *Saturn* in the center of the orbs of his five secondary planets and *Jupiter* in the center of his four secondary planets, and the earth in the center of the moon's orb; and therefore, had this cause been a blind one without contrivance or design, the sun would have been a body of the same kind with *Saturn, Jupiter,* and the earth, that is, without light and heat. Why there is one body in our system qualified to give light and heat to all the rest, I know no reason, but because the author of the system thought it convenient; and why there is but one body of this kind, I know no reason, but because one was sufficient to warm and enlighten all the rest. For the *Cartesian* hypothesis of suns losing their light, and then turning into comets, and comets into planets, can have no place in my System and is plainly erroneous; because it is certain, that, as often as they appear to us, they descend into the system of our planets, lower than the orb of *Jupiter* and sometimes lower than the orbs of *Venus* and *Mercury,* and yet never stay here, but always return from the sun with the same degrees of motion by which they approached him.

To your second Query, I answer that the motions, which the planets now have, could not spring from any natural cause alone, but were impressed by an intelligent Agent. For since comets descend into the region of our planets, and here move all manner of ways, going sometimes the same way with the planets, sometimes the contrary way, and sometimes in crossways, [in] planes inclined to the plane of the ecliptic and at all kinds of angles, it is plain that there is no natural cause which could determine all the planets, both primary and secondary, to move the same way and in the same plane, without any considerable variation: this must have been the effect of counsel. Nor is there any natural cause which could give the planets those just degrees of velocity, in proportion to their distances from the sun and other central bodies, which were requisite to make them move in such concentric orbs about those bodies. Had the planets been as swift as comets, in proportion to their distances from the sun (as they would have been, had their motion been caused by their gravity, whereby the matter, at the first formation of the planets, might fall from the remotest regions toward the sun), they would not move in concentric orbs, but in such eccentric ones as the comets move in. Were all the planets as swift as *Mercury* or as slow as Saturn or his satellites, or were their several velocities otherwise much greater or less than they are, as they might have been, had they arose from any other cause than their gravities, or had the distances from the centers, about which they move, been greater or less than they are, with the same velocities, or had the quantity of matter in the sun, or in *Saturn, Jupiter,* and the earth, and by consequence their gravitating power been greater or less than it is, the primary planets could not have revolved about the sun, nor the secondary ones about *Saturn, Jupiter,* and the earth, in concentric circles, as they do, but would have moved in hyperbolas or parabolas, or in ellipses very eccentric. To make this system, therefore, with all its motions, required a cause which understood, and compared together, the quantities of matter in the several bodies of the sun and planets and the gravitating powers resulting from thence, the several distances of the primary planets from the sun and of the secondary ones from *Saturn, Jupiter,* and the earth, and the velocities with which these planets could revolve about those quantities of matter in the central bodies; and to compare and adjust all these things together, in so great a variety of bodies, argues that cause to be not blind and fortuitous, but very well skilled in mechanics and geometry.

To your third Query I answer, that it may be represented that the sun may, by heating those planets most of which are nearest to him, cause them to be better concocted, and more condensed by that concoction. But when I consider that our earth is much more heated in its bowels below the upper crust, by subterraneous fermentations of mineral bodies than by the sun, I see not why the interior parts of *Jupiter* and *Saturn* might not be as much heated, concocted, and coagulated by those fermentations as our earth is: and therefore this various density should have some other cause than the various distances of the planets from the sun. And I am confirmed in this opinion by considering, that the planets of *Jupiter* and *Saturn*, as they are rarer than the rest, so they are vastly greater, and contain a far greater quantity of matter, and have many satellites about them; which qualifications surely arose not from their being placed at so great a distance from the sun, but were rather the cause why the Creator placed them at great distance. For, by their gravitating powers they disturb one another's motions very sensibly, as I find by some late observations of Mr. *Flamsteed*; and had they been placed much nearer to the sun and to one another, they would by the same powers have caused a considerable disturbance in the whole system.

To your fourth Query, I answer that, in the hypothesis of vortices, the inclination of the axis of the earth might, in my opinion, be ascribed to the situation of the earth's *vortex* before it was absorbed by the neighboring *vortices* and the earth turned from a sun to a comet; but this inclination ought to decrease constantly in compliance with the motion of the earth's *vortex*, whose axis is much less inclined to the ecliptic, as appears by the motion of the moon carried about therein. If the sun by his rays could carry about the planets, yet I do not see how he could thereby effect their diurnal motions.

Lastly, I see nothing extraordinary in the inclination of the earth's axis for proving a Deity, unless you will urge it as a contrivance for winter and summer, and for making the earth habitable toward the poles; and that the diurnal rotations of the sun and planets, as they could hardly arise from any cause purely mechanical, so by being determined all the same way with the annual and menstrual motions, they seem to make up that harmony in the system which, as I explained above, was the effect of choice rather than chance.

There is yet another argument for a Deity, which I take to be a very strong one; but till the principles on which it is grounded are better received, I think it more advisable to let it sleep.

I am, &c.

[Isaac Newton]

Cambridge, Dec. 10, 1692

LETTER II

Sir,

I agree with you that if matter, evenly diffused through a finite space, not spherical, should fall into a solid mass, this mass would affect the figure of the whole space, provided it were not soft, like the old chaos, but so hard and solid from the beginning that the weight of its protuberant parts could not make it yield to their pressure. Yet by earthquakes loosening the parts of this solid, the protuberances might sometimes sink a little by their weight, and thereby the mass might, by degrees, approach a spherical figure.

The reason why matter evenly scattered through a finite space would convene in the midst, you conceive the same with me, but that there should be a central particle, so accurately placed in the middle, as to be always equally attracted on all sides, and thereby continue without motion, seems to me a supposition fully as hard as to make the sharpest needle stand upright on its point upon a looking-glass. For if the very mathematical center of the central particle be not accurately in the very mathematical center of the attractive power of the whole mass, the particle will not be attracted equally on all sides. And much harder it is to suppose all the particles in an infinite space should be so accurately poised one among another, as to stand still in a perfect equilibrium. For I reckon this as hard as to make, not one needle only, but an infinite number of them (so many as there are particles in an infinite space) stand accurately poised upon their points. Yet I grant it possible, at least by a divine power; and if they were once to be placed, I agree with you, that they would continue in that posture without motion forever, unless put into new motion by the same power. When therefore I said that matter, evenly spread through all space. would convene by its gravity into one or more great masses, I understand it of matter not resting in an accurate poise.

But you argue, in the next paragraph of your letter, that every particle of matter in an infinite space has an infinite quantity of matter on all sides and by consequence an infinite attraction every way, and therefore must rest *in equilibrio,* because all infinites are equal. Yet you suspect a paralogism in this argument, and I conceive the paralogism lies in the position that all infinites are equal. The generality of mankind consider infinites no other ways than indefinitely; and in this sense they say all infinites are equal, though they would speak more truly if they should say, they are neither equal nor unequal, nor have any certain difference, or proportion one to another. In this sense, therefore, no conclusions can be drawn from them about the equality, proportions, or differences of things; and they that attempt to do it, usually fall into paralogisms. So when men argue against the infinite divisibility of magnitude, by saying that if an inch may be divided into an infinite number of parts, the sum of those parts will be an inch; and if a foot may be divided into an infinite number of parts, the sum of those parts, must be a foot; and therefore, since all infinites are equal, those sums must be equal, that is, an inch equal to a foot.

The falseness of the conclusion shows an error in the premises, and the error lies in the position that all infinites are equal. There is therefore another way of considering infinites, used by mathematicians, and that is under certain definite restrictions and limitations, whereby infinites are determined to have certain differences or proportions to one another. Thus Dr. *Wallis* considers them in his *Arithmetica Infinitorum,* where, by the various proportions of infinite sums, he gathers the various proportions of infinite magnitudes, which way of arguing is generally allowed by mathematicians, and yet would not be good, were all infinites equal. According to the same way of considering infinites, a mathematician would tell you that though there be an infinite number of infinite little parts in an inch, yet there is twelve times that number of such parts in a foot; that is, the infinite number of those parts in a foot is not equal to but twelve times bigger than the infinite number of them in an inch. And so a mathematician will tell you that if a body stood *in equilibrio* between any two equal and contrary attracting infinite forces, and if to either of these forces you add any new finite attracting force, that new force, howsoever little soever, will destroy their equilibrium, and put the body into the same motion into which it would put it, were those two contrary equal forces but finite, or even none at all; so that in this case the two equal infinites, by the addition of a finite to either of them, become unequal in our ways of reckoning; and after these ways we must reckon, if from the considerations of infinites we would always draw true conclusions.

To the last part of your letter I answer, first, that if the earth (without the moon) were placed anywhere with its center in the *Orbis Magnus,*[84] and stood still there without any gravitation or projection, and there at once were infused into it both a gravitating energy toward the sun and a transverse impulse of a just quantity moving it directly in a tangent to the *Orbis Magnus,* the compounds of this attraction and projection would, according to my notion, cause a circular revolution of the earth about the sun. But the transverse impulse must be a just quantity; for if it be too big or too little, it will cause the earth to move in some other line. Secondly, I do not know any power in Nature which would cause this transverse motion without the Divine arm. *Blondel* tells us somewhere in his book of Bombs that *Plato* affirms that the motion of the planets is such, as if they had all of them been created by God in some region very remote from our system, and let fall from thence toward the sun, and so soon as they arrived at their several orbs their motion of falling turned aside into a transverse one. And this is true, supposing the gravitating power of the sun was double at that moment of time in which they all arrive at their several

---

[84] By *Orbis Magnus*, Newton means the orbit of the Earth. [MJC]

orbs; but then the Divine power is here required in a double respect, namely, to turn the descending motions of the falling planets into a side motion and, at the same time, to double the attractive power of the sun. So then gravity may put the planets into motion, but without the Divine power it could never put them into such a circulating motion as they have about the sun; and therefore, for this as well as other reasons, I am compelled to ascribe the frame of this system to an intelligent Agent.

You sometimes speak of gravity as essential and inherent to matter. Pray do not ascribe that notion to me, for the cause of gravity is what I do not pretend to know, and therefore would take more time to consider of it.

I fear what I have said of infinites will seem obscure to you; but it is enough if you understand that infinites, when considered absolutely without any restriction or limitation, are neither equal nor unequal, nor have any certain proportion one to another, and therefore the principle that all infinites are equal is a precarious one.

I am, Sir, &c.

[Isaac Newton]

Trinity College

Jan. 1, 1692/3

LETTER [III][85]

Sir,

The hypothesis of deriving the frame of the world by mechanical principles from matter evenly spread through the heavens being inconsistent with my system, I had considered it very little before your letters put me upon it, and therefore trouble you with a line or two more about it, if this comes not too late for your use.

In my former I represented that the diurnal rotations of the planets could not be derived from gravity, but required a Divine Arm to impress them. And though gravity might give the planets a motion of descent toward the sun, either directly or with some little obliquity, yet the transverse motions by which they revolve in their several orbs required the divine arm to impress them according to the tangents of their orbs. I would now add that the hypothesis of matters being at first evenly spread through the heavens is, in my opinion, inconsistent with the hypothesis of innate gravity, without a supernatural power

---

[85] Although the editor of this edition of Newton's works labeled this letter IV, it is dated earlier than Letter III in his numeration. I have restored their positions so as to preserve chronological order. [MJC]

to reconcile them; and therefore it infers a Deity. For if there be innate gravity, it is impossible now for the matter of the earth and all the planets and stars to fly up from them, and become evenly spread throughout all the heavens, without a supernatural power; and certainly that which can never be hereafter without a supernatural power, could never be heretofore without the same power.

You queried whether matter evenly spread throughout a finite space, of some other figure than spherical, would not, in falling down toward a central body, cause that body to be of the same figure with the whole space, and I answered, yes. But in my answer it is to be supposed that the matter descends directly downward to that body, and that that body has no diurnal rotation.

This, sir, is all I would add to my former letters.

I am yours, &c

[Isaac Newton]

Cambridge,

Feb. 11, 1693

LETTER [IV]

Sir,

Because you desire speed, I will answer your letter with what brevity I can. In the six positions you lay down in the beginning of your letter, I agree with you. Your assuming the *Orbis Magnus* 7,000 diameters of the earth wide implies the sun's horizontal parallax to be half a minute. *Flamsteed* and *Cassini* have of late observed it to be about 10″, and thus the *Orbis Magnus* must be 21,000, or in a round number 20,000 diameters of the earth wide. Either computation I think will do well, and I think it not worth while to alter your numbers.

In the next part of your letter you lay down four other positions founded upon the six first. The first of these four seems very evident, supposing you take attraction so generally as by it to understand any force by which distant bodies endeavor to come together without mechanical impulse. The second seems not so clear. For it may be said that there might be other systems of worlds before the present ones, and others before those, and so on to all past eternity, and by consequence that gravity may be co-eternal to matter and have the same effect from all eternity as at present, unless you have somewhere proved that old systems cannot gradually pass into new ones or that this system had not its original from the exhaling matter of former decaying systems but from a chaos of matter evenly dispersed throughout all space. For something of this kind, I think you say, was the subject of your Sixth Sermon, and the growth of new systems out of old ones, without the mediation of a Divine Power, seems to me apparently absurd.

The last clause of the second position I like very well. It is inconceivable that inanimate brute matter should, without the mediation of something else which is not material, operate upon and affect other matter without mutual contact, as it must be if gravitation, in the sense of *Epicurus*, be essential and inherent in it. And this is one reason why I desired you would not ascribe innate gravity to me. That gravity should be innate, inherent, and essential to matter, so that one body may act upon another at a distance through a *vacuum*, without the mediation of any thing else, by and through which their action and force may be conveyed from one to another, is to me so great an absurdity that I believe no man who has in philosophical matters a competent faculty of thinking can ever fall into it. Gravity must be caused by an agent acting constantly according to certain laws, but whether this agent be material or immaterial I have left to the consideration of my readers. Your fourth assertion, that the world could not be formed by innate gravity alone, you confirm by three arguments. But in your first argument you seem to make a *Petitio Principii*; for whereas many ancient philosophers and others, as well Theists as Atheists, have all allowed that there may be worlds and parcels of matter innumerable or infinite, you deny this by representing it as absurd as that there should be positively an infinite arithmetical sum or number, which is a contradiction *in terminis*, but you do not prove it as absurd. Neither do you prove that what men mean by an infinite sum or number is a contradiction in nature, for a contradiction *in terminis* implies no more than an impropriety of speech. Those things, which men understand by improper and contradictious phrases, may be sometimes really in Nature without any contradiction at all: a silver inkhorn, a paper lantern, an iron whetstone, are absurd phrases, yet the things signified thereby are really in Nature. If any man should say that a number and a sum, to speak properly, is that which may be numbered and summed, but things infinite are numberless or, as we usually speak, innumerable and sumless or insummable, and therefore ought not to be called a number or sum, he will speak properly enough, and your argument against him will, I fear, lose its force. And yet if any man shall take the words 'number' and 'sum' in a larger sense, so as to understand thereby things which, in the proper way of speaking, are numberless and sumless (as you seem to do when you allow an infinite number of points in a line), I could readily allow him the use of the contradictious phrases of 'innumerable number' or 'sumless sum,' without inferring from thence any absurdity in the thing he means by those phrases. However, if by this or any other argument you have proved the finiteness of the universe, it follows that all matter would fall down from the outsides and convene in the middle. Yet the matter in falling might concrete into many round masses, like the bodies of the planets, and these, by attracting one another, might acquire an obliquity of descent by means of which they might fall, not upon the great central body, but upon the side of it, and fetch a compass about and then ascend again by the same steps and degrees of motion and velocity with which they descended before, much after the manner that comets revolve about the sun; but a circular motion in concentric orbs about the sun, they could never acquire by gravity alone.

And though all the matter were divided at first into several systems, and every system by a Divine Power constituted like ours, yet would the outside systems descend toward the middlemost; so that this frame of things could not always subsist without a Divine Power to conserve it, which is the second argument; and to your third I fully assent.

As for the passage of *Plato*, there is no common place from whence all the planets, being let fall and descending with uniform and equal gravities (as *Galileo* supposes), would, at their arrival to their several orbs, acquire their several velocities with which they now revolve in them. If we suppose the gravity of all the planets toward the sun to be of such a quantity as it really is, and that the motions of the planets are turned upward, every planet will ascend to twice its height from the sun. *Saturn* will ascend till he be twice as high from the sun as he is at present, and no higher; *Jupiter* will ascend as high again as at present, that is, a little above the orb of *Saturn*; *Mercury* will ascend to twice his present height, that is, to the orb of *Venus*; and so of the rest; and then, by falling down again from the places to which they ascended, they will arrive again at their several orbs with the same velocities they had at first and with which they now revolve.

But if, so soon as their motions by which they revolve are turned upward, the gravitating power of the sun, by which their ascent is perpetually retarded, be diminished by one half, they will now ascend perpetually, and all of them at equal distances from the sun will be equally swift. *Mercury*, when he arrives at the orb of *Venus*, will be as swift as *Venus*; and he and Venus, when they arrive at the orb of the earth, will be as swift as the earth; and so of the rest. If they begin all of them to ascend at once and ascend in the same line, they will constantly, in ascending, become nearer and nearer together, and their motions will constantly approach to an equality and become at length slower than any motion assignable. Suppose, therefore, that they ascended till they were almost contiguous and their motions inconsiderably little, and that all their motions were at the same moment of time turned back again or, which comes almost to the same thing, that they were only deprived of their motions and let fall at that time; they would all at once arrive at their several orbs, each with the gravitating power of the sun doubled, that it might be strong enough to retain them in their orbs, they would revolve in them as before their ascent. But if the gravitating power of the sun was not doubled, they would go away from their orbs into the highest heavens in parabolical lines. These things follow from my *Princip[ia] Math[ematica]*, [Book] I, Prop[ositions] XXXIII, XXXIV, XXXV, XXXVI.

I thank you very kindly for your designed present, and rest yours &c

[Isaac Newton]
Cambridge,
Feb. 25, 1692/3

Selections from the Leibniz-Clarke Correspondence

Introduction

Does God supernaturally intervene in the operations of the universe? If so, under what conditions? If not, what does this imply about God's nature? These important questions were widely debated during the eighteenth century. Among the classic discussions of these questions, one of the most famous appeared as part of what is known as the "Leibniz-Clarke Correspondence," which consisted of five pairs of letters exchanged between the brilliant philosopher, mathematician, and historian Gottfried Wilhelm Leibniz (1646–1716) and Rev. Samuel Clarke (1675–1729). One of Leibniz's most important achievements was the development of some central ideas in the calculus, an area of mathematics in which his achievements were surpassed only by those of Newton. In fact, an acrimonious priority dispute had broken out in 1713 between the supporters of Leibniz and Newton. Clarke was a former student and close friend of Newton, who had chosen Clarke to translate his *Opticks* (1704) into Latin. Clarke had also served for a time as chaplain to Queen Anne, through which position he came into close contact with the British Royal Family, including Caroline of Anspach, the Princess of Wales.

In 1715, Leibniz wrote a letter to Princess Caroline in which he criticized Newton's Natural Theology. Portions of this letter form the first letter in the correspondence. Princess Caroline turned to Clarke for a defense of Newton's ideas. It is essentially certain, although no surviving documents directly demonstrate this point, that Clarke consulted Newton in formulating his response. Certainly the statements in Clarke's letters were in agreement with what is known of Newton's ideas. By the time of Leibniz's death on 14 November 1716, five pairs of letters had been exchanged, treating such topics as the existence of a vacuum, whether the universe functions as a sort of self-regulating clockwork, and the nature of space, time, and gravity. Clarke published this collection of letters in 1717 under the title: *A Collection of Papers, Which Passed between the Late Learned Mr. Leibnitz, and Dr. Clarke in the Years 1715 and 1716, Relating to the Principles of Natural Philosophy and Religion.*

The Leibniz-Clarke Correspondence[86]

MR. LEIBNIZ'S FIRST PAPER
being
an Extract of a Letter Written in November, 1715

1. Natural religion itself seems to decay (in England) very much. Many will have human souls to be material; other make God himself a corporeal being.

2. Mr. Locke, and his followers, are uncertain at least whether the soul be not material and naturally perishable.

3. Sir Isaac Newton says that space is an organ, which God makes use of to perceive things by.[87] But if God stands in need of any organ to perceive things by, it will follow that they do not depend altogether upon him, nor were produced by him.

4. Sir Isaac Newton, and his followers, have also a very odd opinion concerning the work of God. According to their doctrine, God

---

[86] From *A Collection of Papers, Which Passed between the Late Learned Mr. Leibnitz, and Dr. Clarke in the Years 1715 and 1716, Relating to the Principles of Natural Philosophy and Religion* as printed in *The Works of Samuel Clarke*, vol. 4 (John and Paul Knapton, 1738), pp. 587–588, 590–591, 593–596, 598–601, 605–606, and 610–612. Note: Spelling, punctuation, capitalization, and such matters have been slightly modernized for this printing. Included in this change is the spelling of Leibniz's name as Leibniz, rather than Leibnitz.

[87] It was long believed that this was an unfair accusation for Leibniz to bring against Newton. Then it was discovered that there is an edition of Newton's *Opticks* in some copies of which Newton had stated: "Is not the whole of Space the Sensorium of a Being incorporeal, living, and Intelligent; in that he sees distinctly and closely comprehends the most inward things themselves..." Newton, however, had apparently decided this was too strong a statement and replaced it by a much weaker statement in which he compared space with the sensorium of God. On this issue, see I. B. Cohen and A. Koyré, "The Case of the Missing *Tanquam*," *Isis, 52* (1961), 555–566 and F. E. L. Priestley, "The Clarke-Leibniz Controversy" in Robert Butts and John Davis (eds.), *The Methodological Heritage of Newton* (Basil Blackwell, 1970), pp. 34–56. [MJC]

Almighty wants to wind up his watch from time to time: otherwise it would cease to move.[88] He had not, it seems, sufficient foresight to make it a perpetual motion. Nay, the machine of God's making is so imperfect, according to these gentlemen, that he is obliged to clean it now and then by an extraordinary concourse, and even to mend it, as a clockmaker mends his work; who must consequently be so much the more unskillful a workman, as he is oftener obliged to mend his work and set it right. According to my opinion, the same force and vigour remains always in the world, and only passes from one part of matter to another, agreeably to the laws of nature, and the beautiful pre-established order.[89] And I hold that when God works miracles, he does not do it in order to supply the wants of nature, but those of grace. Whoever thinks otherwise, must needs have a very mean notion of the wisdom and power of God.

DR. CLARKE'S FIRST REPLY

4. The reason why, among men, an artificer is justly esteemed so much the more skillful, as the machine of his composing will continue longer to move regularly without any farther interposition of the workman, is because the skill of all human artificers consists only in composing, adjusting, or putting together certain movements, the principles of whose motion are altogether independent upon the artificer: such as are weights and springs, and the like, whose forces are not made, but only adjusted, by the workman. But with regard to God, the case is quite different, because he not only composes or puts things together, but is himself the author and continual preserver of their original forces or moving powers: and consequently 'tis not a diminution, but the true glory of his workmanship, that nothing is done without his continual government and inspection. The notion of the world's being a great machine, going on without the interposition of God, as a clock continues to go without the assistance of a

---

[88] Leibniz was very possibly referring to a statement made by Newton in the 28th Query in his *Opticks*, where Newton had written: "For while comets move in very eccentric orbs in all manner of positions, blind fate could never make all the planets move one and the same way in orbs concentric, some inconsiderable irregularities excepted which may have arisen from the mutual actions of comets and planets upon one another, and which will be apt to increase, till this system wants a reformation." [MJC]

[89] Put in slightly more technical language, Leibniz is maintaining that God gave the universe at its creation a certain quantity of motion, which quantity is conserved. Note that Leibniz favored the view that this conserved quantity of motion was not mv, as Descartes seemed to believe, but rather mv^2, or what is now known as kinetic energy, i.e., $\frac{mv^2}{2}$. [MJC]

clockmaker, is the notion of materialism and fate, and tends (under pretense of making God a *supra-mundane intelligence*) to exclude providence and God's government in reality out of the world. And by the same reason that a philosopher can represent all things going on from the beginning of the creation, without any government or interposition of providence, a sceptic will easily argue still farther backwards, and suppose that things have from eternity gone on (as they now do) without any true creation or original author at all, but only what such arguers call all-wise and eternal nature. If a king had a Kingdom, wherein all things would continually go on without his government or interposition, or without his attending to and ordering what is done therein, it would be to him, merely a nominal kingdom; nor would he in reality deserve at all the title of king or governor. And as those men, who pretend that in an earthly government things may go on perfectly well without the king himself ordering or disposing of anything, may reasonably be suspected that they would like very well to set the king aside: so whosoever contends, that the course of the world can go on without the continual direction of God, the Supreme Governor, his doctrine does in effect tend to exclude God out of the world.

MR. LEIBNIZ'S SECOND PAPER
being
an Answer to Dr. Clarke's First Reply

6. The true and principal reason why we commend a machine is rather grounded upon the effect of the machine than upon its cause. We don't enquire so much about the power of the artist, as we do about his skill in his workmanship. And therefore the reason alleged by the author for extolling the machine of God's making, grounded upon his having made it entirely, without wanting any materials to make it of; that reason, I say, is not sufficient. 'Tis a mere shift the author has been forced to have recourse to: and the reason why God exceeds any other artist is not only because he makes the whole, whereas all other artists must have matter to work upon. This excellency in God would be only on the account of power. But God's excellency arises also from another cause, viz. wisdom: whereby his machine lasts longer, and moves more regularly, than those of any other artist whatsoever. He who buys a watch, does not mind whether the workman made every part of it himself, or whether he got the several parts made by others, and did only put them together, provided the watch goes right. And if the workman had received from God even the gift of creating the matter of the wheels, yet the buyer of the watch

would not be satisfied, unless the workman had also received the gift of putting them well together. In like manner, he who will be pleased with God's workmanship cannot be so, without some other reason than that which the author has here alleged.

7. Thus the skill of God must not be inferior to that of a workman; nay, it must go infinitely beyond it. The bare production of every thing would indeed show the *power* of God, but it would not sufficiently show his *wisdom.* They who maintain the contrary will fall exactly into the error of the materialists, and of Spinoza, from whom they profess to differ. They would, in such case, acknowledge *power,* but not sufficient *wisdom,* in the principle or cause of all things.

8. I do not say the material world is a machine, or watch, that goes without God's interposition; and I have sufficiently insisted, that the creation wants to be continually influenc'd by its creator. But I maintain it to be a watch, that goes without wanting to be mended by him: otherwise we must say, that God bethinks himself again. No; God has foreseen every thing; he has provided a remedy for every thing before-hand; there is in his works a harmony, a beauty, already pre-established.

9. This opinion does not exclude God's providence, or his government of the world; on the contrary, it makes it perfect. A true providence of God requires a perfect foresight. But then it requires, moreover, not only that he should have foreseen every thing; but also that he should have provided for every thing beforehand, with proper remedies; otherwise, he must want either *wisdom* to foresee things, or *power* to provide against them. He will be like the God of the Socinians, who lives only from day to day, as Mr. Jurieu says. Indeed God, according to the Socinians, does not so much as foresee inconveniences; whereas, the gentlemen I am arguing with, who put him upon mending his work, say only, that he does not provide against them. But this seems to me to be still a very great imperfection. According to this doctrine, God must want either power, or good will.

10. I don't think I can be rightly blamed, for saying that God is *intelligentia supramundana.* Will they say, that he is *intelligentia mundana*; that is, the soul of the world? I hope not. However, they will do well to take care, not to fall into that notion unawares.

11. The comparison of a king, under whose reign every thing should go on without his interposition, is by no means to the present purpose, since God preserves every thing continually, and nothing can subsist without him. His kingdom therefore is not a nominal one. 'Tis just as if one should say that a king, who should originally have taken care to have his subjects so well educated, and should, by his

care in providing for their subsistence, preserve them so well in their fitness for their several stations, and in their good affection towards him, as that he should have no occasion ever to be mending any thing amongst them, would be only a nominal king.

12. To conclude. If God is oblig'd to mend the course of nature from time to time, it must be done either supernaturally or naturally. If it is done supernaturally, we must have recourse to miracles, in order to explain natural things, which is reducing an hypothesis *ad absurdum*, for, every thing may easily be accounted for by miracles. But if it be done naturally, then God will not be *intelligentia supramundana*; he will be comprehended under the nature of things; that is, he will be the soul of the world.

DR. CLARKE'S SECOND REPLY

6 and 7. 'Tis very true that the excellency of God's workmanship does not consist in its showing the *power* only, but in its showing the *wisdom* also of its author. But then this *wisdom* of God appears, not in making nature (as an artificer makes a clock) capable of going without him: (for that's impossible, there being no powers of nature independent upon God, as the powers of weights and springs are independent upon men), but the *wisdom* of God consists, in framing originally the perfect and complete idea of a work, which begun and continues, according to that original perfect idea, by the continual uninterrupted exercise of his *power* and *government*.

8. The word *correction*, or *amendment*, is to be understood, not with regard to God, but to us only. The present frame of the solar system (for instance) according to the present laws of motion, will in time fall into confusion; and perhaps, after that, will be amended or put into a new form. But this amendment is only relative with regard to our conceptions. In reality, and with regard to God; the present frame, and the consequent disorder, and the following renovation, are all equally parts of the design framed in God's original perfect idea. 'Tis in the frame of the world, as in the frame of man's body: the wisdom of God does not consist in making the present frame of either of them eternal, but to last so long as he thought fit.

9. The *wisdom* and *foresight* of God, do not consist in providing original remedies, which shall of themselves cure the disorders of nature. For in truth and strictness, with regard to God, there are no disorders, and consequently no remedies; and indeed no powers of nature at all, that can do any thing of themselves (as weights and springs work of themselves with regard to men); but the wisdom and

foresight of God consist (as has been said) in contriving at once, what his power and government is continually putting in actual execution.

10. God is neither a *mundane intelligence,* nor a *supramundane intelligence*; but an omnipresent intelligence, both in and without the world. He is in all, and through all, as well as above all.

11. If God's conserving all things means his actual operation and government, in preserving and continuing the beings, powers, orders, dispositions and motions of all things, this is all that is contended for. But if his conserving things, means no more than a king's creating such subjects, as shall be able to act well enough without his intermeddling or ordering any thing amongst them ever after; this is making him indeed a real creator, but a governor only nominal.

12. The argument in this paragraph supposes that whatsoever God does is supernatural or miraculous; and consequently it tends to exclude all operation of God in the governing and ordering of the natural world. But the truth is: *natural* and *supernatural* are nothing at all different with regard to God, but distinctions merely in our conceptions of things. To cause the sun (or earth) to move regularly is a thing we call natural; to stop its motion for a day, we call supernatural; but the one is the effect of no greater power, than the other; nor is the one, with respect to God more or less natural or supernatural than the other. God's being present in or to the world, does not make him to be the soul of the world. A soul is part of a compound, whereof body is the other part; and they mutually affect each other, as parts of the same whole. But God is present to the world, not as a part, but as governor; acting upon all things, himself acted upon by nothing. He is not far from every one of us, for in him we (and all things) live and move and have our beings.

MR. LEIBNIZ'S THIRD PAPER
being
an Answer to Dr. Clarke's Second Reply

14. When I said that God has provided remedies before-hand against such disorders, I did not say that God suffers disorders to happen, and then finds remedies for them, but that he has found a way before-hand to prevent any disorders happening.

15. The author strives in vain to criticize my expression that God is *intelligentia supramundana*. To say that God is above the world is not denying that he is in the world.

16. I never gave any occasion to doubt, but that God's conservation is an actual preservation and continuation of the beings, powers,

orders, dispositions, and motions of all things: and I think I have perhaps explained it better than many others. But, says the author, *this is all that I contended for*. To this I answer; *your humble servant for that, sir*. Our dispute consists in many other things. The question is, whether God does not act in the most regular and most perfect manner? Whether his machine is liable to disorders, which he is obliged to mend by extraordinary means? Whether the will of God can act without reason? Whether space is an absolute being? Also concerning the nature of miracles; and many such things, which make a wide difference between us.

17. Divines will not grant the author's position against me, viz. that there is no difference, with respect to God, between *natural* and *supernatural*: and it will be still less approved by most philosophers. There is a vast difference between these two things, but it plainly appears, it has not been duly consider'd. That which is supernatural, exceeds all the powers of creatures. I shall give an instance, which I have often made use of with good success. If God would cause a body to move free in the aether round about a certain fixed centre, without any other creature acting upon it, I say, it could not be done without a miracle, since it cannot be explained by the nature of bodies. For, a free body does naturally recede from a curve in a tangent. And therefore I maintain that the attraction of bodies, properly so called, is a miraculous thing, since it cannot be explained by the nature of bodies.

DR. CLARKE'S THIRD REPLY

13. and 14. The *active forces*, which are in the universe, diminishing themselves so as to stand in need of new impressions, is no inconvenience, no disorder, no imperfection in the workmanship of the universe, but is the consequence of the nature of dependent things. Which dependency of things is not a matter that wants to be rectified. The case of a human workman making a machine, is quite another thing, because the powers or forces by which the machine continues to move are altogether independent on the artificer.

15. The phrase, *intellegentia supramundana*, may well be allowed, as it is here explained, but without this explication, the expression is very apt to lead to a wrong notion, as if God was not really and substantially present every where.

16. To the questions here proposed, the answer is: that God does always act in the most regular and perfect manner; that there is nothing more extraordinary in the alterations he is pleased to make in the frame of things, than in his continuation of it; that in things in their

own nature absolutely equal and indifferent, the will of God can freely choose and determine itself, without any external cause to impel it; and that 'tis a perfection in God to be able so to do; that space, does not at all depend on the order or situation of existence of bodies. And as to the notion of miracles.

17. The question is not, what it is that divines or philosophers usually allow or not allow, but what reasons men allege for their opinions. If a miracle be that only, which surpasses the power of all created beings, then for a man to walk on the water, or for the motion of the sun or the earth to be stopped, is no miracle, since none of these things require infinite power to effect them. For a body to move in a circle round a centre *in vacuo;* if it be usual (as the planets moving about the sun), 'tis no miracle, whether it be effected immediately by God himself, or mediately by any created power: but if it be usual (as, for a heavy body to be suspended, and move so in the air), 'tis a miracle, whether it be effected immediately by God himself, or mediately by any invisible created power. Lastly, if whatever arises not from, and is not explicable by, the natural powers of body, be a miracle, then every animal-motion whatsoever, is a miracle. Which seems demonstrably to show, that this learned author's notion of a miracle is erroneous.

Newton, Voltaire, and Cartesianism

Newtonian ideas gradually gained recognition throughout continental Europe, but frequently not without a struggle against Cartesian doctrines. Some sense of this situation can be gained from a selection from one of the *Lettres philosophiqes* of Voltaire (1694–1778). Early recognized as a person capable of creating delightful writings and also much controversy, Voltaire had by 1722 served an eleven month prison term and had proclaimed his goal of becoming the "new Lucretius." Back in Bastille in 1726, he was freed on the condition that he accept exile. Thus he came to spend two years with the English, whose liberal politics, Lockean philosophy, and Newtonian science he espoused with enthusiasm and soon urged upon his countrymen in his *Lettres philosophiques* (1733).

In his Letter XIV, entitled "On Descartes and Sir Isaac Newton," Voltaire contrasted the universe of the English with that of the French by writing:

> A Frenchman who arrives in London, will find philosophy, like everything else, very much changed there. He had left the world a plenum, and he now finds it a vacuum. At Paris the universe is seen composed of vortices of subtile matter; but nothing like it is seen in London. In France, it is the pressure of the moon that causes the tides; but in England it is the sea that gravitates towards the moon; so that when

you think that the moon should make it flood with us, those gentlemen fancy it should be ebb, which very unluckily cannot be proved. For to be able to do this, it is necessary the moon and the tides should have been inquired into at the very instant of the creation.

You will observe farther, that the sun, which in France is said to have nothing to do in the affair, comes in here for very near a quarter of its assistance. According to your Cartesians, everything is performed by an impulsion, of which we have very little notion; and according to Sir Isaac Newton, it is by an attraction, the cause of which is as much unknown to us. At Paris you imagine that the earth is shaped like a melon, or of an oblique figure; at London it has an oblate one. A Cartesian declares that light exists in the air; but a Newtonian asserts that it comes from the Sun in six minutes and a half. The several operations of your chemistry are performed by acids, alkalies, and subtile matter; but attraction prevails even in chemistry among the English.

The very essence of things is totally changed. . . .[90]

That Voltaire sided with the Newtonians became even more clear in 1738, when he published an exposition of the Newtonian system entitled *Élements de la philosophie de Newton,* which is one of a number of documents that carried Newtonian ideas to the continent. Voltaire's creativity, however, did not lie in science. In fact, his mistress, Madame du Châtelet (1706–1749), was more gifted in mathematics, as is evident from the fact that she is famous for having prepared a French translation of and commentary on Newton's *Principia.*[91]

Some Quotations Concerning Newton

G. W. Leibniz (1646–1716)

[Sir A. Fontaine reported that when the Queen of Prussia asked Leibniz his opinion of Newton], Leibniz said that taking Mathematicks from the beginning to the world to the time of S[ir] Isaac, What he had done was much the better half. . . .[92]

J. L. Lagrange (1736–1813)

Newton was the greatest genius that ever existed, and the most fortunate, for we cannot find more than once a system of the world to establish.[93]

---

[90] As given in Voltaire, *Letters on the English* in French and English Philosophers, vol. 38 of the Harvard Classics (P. F. Collier and Son, 1910), pp. 108–109.

[91] On Madame du Châtelet, see Carolyn Iltis, "Madame du Châtelet's Metaphysics and Mechanics," *Studies in the History of Physical Sciences, 8* (1977), 29–48.

[92] As quoted in Richard S. Westfall, *Never at Rest,* p. 721.

[93] As quoted in F. R. Moulton, *Introduction to Astronomy* (Macmillan, 1906), p. 199.

Samuel Taylor Coleridge (1772–1834)

The more I understand of Sir Isaac Newton's works, the more boldly I dare utter to my own mind, & therefore to *you,* that I believe the Souls of 500 Sir Isaac Newtons would go to the making up of a Shakespeare or a Milton.[94]

William Wordsworth (1770–1850)

> The Evangelist St John my patron was:
> Three Gothic courts are his, and in the first
> Was my abiding-place, a nook obscure;
> Right underneath, the College kitchens made
> A humming sound, less tuneable than bees,
> But hardly less industrious; with shrill notes
> Of sharp command and scolding intermixed.
> Near me hung Trinity's loquacious clock,
> Who never let the quarters, night or day,
> Slip by him unproclaimed, and told the hours
> Twice over with a male and female voice.
> His pealing organ was my neighbour too;
> And from my pillow, looking forth by light
> Of moon or favouring stars, I could behold
> The antechapel where the statue stood
> Of Newton with his prism and silent face,
> The Marble index of a mind for ever
> Voyaging through strange seas of Thought, alone.[95]

David Brewster (1781–1868)

The name of Sir Isaac Newton has by general consent been placed at the head of those great men who have been the ornaments of their species. . . . The philosopher [Laplace], indeed, to whom posterity will probably assign a place next to Newton, has characterized the *Principia* as preeminent above all the productions of human intellect.[96]

William Whewell (1794–1866)

[Newton's discovery of the law of universal gravitation] is indisputably and incomparably the greatest scientific discovery ever made, whether we look at the advance which it involved, the extent of truth disclosed, or the fundamental and satisfactory nature of the truth.[97]

---

[94] In an 1801 letter quoted in A. J. Meadows, *The High Firmament* (Leicester University Press, 1969), p. 163.

[95] *The Prelude,* Book III, lines 46–63.

[96] David Brewster, *Life of Sir Isaac Newton* (J. Murray, 1831), pp. 1, 2.

[97] William Whewell, *History of the Inductive Sciences,* vol. 2 (John W. Parker, 1837), p. 180.

Newton could not admit that there was any difference between him and other men, except in the possession of such habits as. . . perseverance and vigilance. When he was asked how he made his discoveries, he answered, "by always thinking about them;" and at another time he declared that if he had done anything, it was due to nothing but industry and patient thought: "I keep the subject of my inquiry constantly before me, and wait till the first dawning opens gradually, by little and little, into a full and clear light."[98]

Albert Einstein (1879–1955)

Fortunate Newton, happy childhood of science! He who has time and tranquillity can by reading this book live again the wonderful events which the great Newton experienced in his young days. Nature was to him an open book, whose letters he could read without effort. The conceptions which he used to reduce material of experience to order seemed to flow spontaneously from experience itself, from the beautiful experiments which he ranged in order like playthings and describes with an affectionate wealth of detail. In one person he combined the experimenter, the theorist, the mechanic, and, not least, the artist in exposition. He stands before us strong, certain, and alone: his joy in creation and his minute precision are evident in every word and in every figure.[99]

---

[98]Whewell, *History of the Inductive Sciences,* vol. 2, p. 185.
[99]Albert Einstein, "Foreword" to Isaac Newton, *Opticks,* 4th ed. (Dover, 1952), p. lix.

Chapter 5

Between Newton and Einstein

Introduction

Physics developed in a number of important and exciting ways in the period between Newton and Einstein. The goal of this chapter will be to survey some of those developments, particularly those that were in the background of Einstein's formulation of his special theory of relativity. One way of pointing to the importance of a number of these developments is to call into question the widespread belief that the two great modern revolutions in physics were (1) that associated especially with Galileo and Newton and (2) that associated with Planck and Einstein. Planck played a key role in the second revolution by his foundational contributions to quantum theory whereas Einstein made seminal contributions to both relativity and quantum theory. Although those two revolutions were profoundly important, a plausible case can be made that three other revolutionary developments occurred in physics in the period between Newton and Einstein.

The following statement put forth in 1938 makes a case for the importance of the first of these three, that concerning the notion of an electromagnetic field, especially the ideas developed in the nineteenth century by Michael Faraday and James Clerk Maxwell.

> A new concept appears in physics, the most important invention since Newton's time: the field. It needed great scientific imagination to realize that it is not the charges nor the particles but the field in the space between the charges and the particles which is essential for the description of physical phenomena.[1]

The plausibility of this claim for the importance of the field concept (which

---

[1] Albert Einstein and Leopold Infeld, *The Evolution of Physics* (Simon and Schuster, 1961 reprint of the 1938 original), p. 244.

will be discussed further below) becomes evident when it is noted that the claim was made in a history of physics published by Albert Einstein himself, writing in collaboration with the physicist Leopold Infeld.

The second candidate for a revolutionary development in the period between Newton and Einstein is the discovery and development of the concept of energy (see below). Persons who have studied physics sometimes find it to be a remarkable fact that before the 1840s when Mayer, Joule, Helmholtz, and Colding developed the law of the conservation of energy, scientists lacked the concept of the entity now so crucial for understanding many aspects of science: the idea of energy. The claim that there exists an invisible entity called energy that manifests itself in mechanical, electrical, chemical, and other forms influenced all aspects of science.

The third candidate for a revolution in physics in the period between Newton and Einstein is allied to the second. It is that in a very meaningful sense, before the creation and elaboration of the notion of energy, physics itself as a unified field of study did not exist. In other words, before the 1840s, such areas as mechanics, light, heat, electricity, magnetism, sound, etc. existed for the most part in isolation from each other. When, however, scientists began to see how to trace the transformation of mechanical energy into, for example, thermal or electrical energy, then physics as a unified discipline emerged. It is not that some of these areas, such as mechanics and optics, had not existed for centuries as areas of inquiry, indeed of very successful inquiry; it is rather that scientists saw them as separate disciplines that could be advanced without detailed consideration of other such disciplines. Partial evidence for the surprising claim that the creation of physics as we know it should be dated from the mid-nineteenth century comes from comparing the founding dates for some of Britain's major scientific societies: whereas the founding of Britain's Geological Society occurred in 1807, of the Astronomical Society in 1820, and of the Chemical Society in 1841, it was only in 1874 that the Physical Society made its appearance. Moreover, the term *physicist* was introduced into English only in 1840. This occurred in a paragraph of William Whewell's *Philosophy of the Inductive Sciences,* where he wrote: "The terminations *ize* (rather than *ise*), *ism,* and *ist,* are applied to words of all origins; thus we have to *pulverize,* to *colonize, Witticism, Heathenism, Journalist, Tobacconist.* Hence we may make such words when they are wanted. As we cannot use *physician* for a cultivator of physics, I have called him a *Physicist.*"[2] Whewell's proposal met a good deal of opposition. For example, Michael Faraday protested: "Physicist is both to my mouth and

---

[2] As quoted in Sydney Ross, "*Scientist*: The Story of a Word," *Annals of Science, 18* (1962), 65–85: p. 72 from Whewell's *Philosophy of the Inductive Sciences,* vol. 1 (John W. Parker, 1840), p. cxiii.

ears so awkward that I think I shall never be able to use it. The equivalent of the three separate sounds of i in one word is too much."[3]

This third candidate for revolutionary status has particular significance for the form and contents of this chapter, which is aimed not at describing all the important advances in physical science in the eighteenth and nineteenth centuries, but rather at setting the stage for Einstein's introduction of the special theory of relativity. What is particularly significant is that whereas the subject matter of Newton's *Principia* is mechanics, Einstein's special theory of relativity had direct implications for, and arose partly from, a number of areas of physics, not just mechanics. Properly to understand Einstein's special theory requires some knowledge of developments during the eighteenth and nineteenth centuries in a variety of areas in physics. To a discussion of these we now proceed.

Part I: Developments in Physics during the Eighteenth and Nineteenth Centuries Relevant to Einstein's Special Theory of Relativity [4]

Mechanics in the Seventeenth Century

First, a brief review. In 1638, **Galileo Galilei** (1564–1642) published his *Discourses and Demonstrations Concerning Two New Sciences,* in which he presented his laws governing the motions of terrestrial bodies. These include the claims that the distance through which a body falls from rest is proportional to the square of its time of fall and that the path of projectiles on the Earth is parabolic. In his *Principia philosophiae* of 1644, **René Descartes** (1596–1650) presented his law of inertia and asserted that the quantity of motion in the universe is a constant. He identified the quantity of motion in a body as proportional to its velocity and its size. This quantity is now known as momentum.

In 1687, **Newton** (1642–1727) published his *Principia*, in which he presented his famous **three laws of motion** as well as the **law of gravitation**. He also provided empirical evidence that the **inertial mass** of a body (that property by which the body resists changes in motion) is numerically equal to its **gravitational mass** (that property on which the gravitational forces on the body depend). Newton used this equality to explain why terrestrial

---

[3]As quoted in Ross, *"Scientist,"* p. 72.

[4]The presentation in this section is based to some extent on the presentation in Einstein and Infeld, *Evolution of Physics.* Readers of that book should be aware that in many cases it provides out of date and even erroneous information regarding particular developments. It does, however, provide insights regarding the development of Einstein's thought.

bodies of differing masses fall with the same acceleration. Put in terms of equations, his explanation is as follows.

1. By the law of gravitation, the gravitational force F on a body is equal to $\frac{Gm_g m_g'}{D^2}$, where m_g is the gravitational mass of the body, m_g' is the gravitational mass of the Earth, D is the body's distance from the Earth, and G is a constant.

2. By Newton's second law, $F = km_i a$, where m_i is the inertial mass of the body, a is the acceleration produced by the force F, and k is a constant.

3. If $m_g = m_i$ (i.e., if a body's gravitational mass is equal to its inertial mass), then from $km_i a = \frac{Gm_g m_g'}{D^2}$, we get $a = \frac{Gm_g'}{kD^2}$, which states that all bodies at distance D will experience the same amount of acceleration. As Einstein and Infeld put it, the numerical equivalence of the gravitational and inertial masses of a body is, in Newton's system, an "accidental" result,[5] that is, Newton did not offer an explanation as to why gravitational and inertial mass should be equal.

In his *Principia,* Newton also held that **absolute space and time** are meaningful concepts.

Mechanics from Newton to Einstein

Despite Alexander Pope's claim that "Nature, and Nature's Laws lay hid in Night:/God said, *Let Newton be!* and All was *Light,*"[6] the Newtonian system only gradually received the recognition that it deserved. In fact, the process by which Newton's mechanics attained almost universal acceptance extended over a number of decades and involved its confronting and successfully meeting various challenges. The most resistance came from Cartesian, typically Continental, scientists. Thomas Hankins in his *Science and the Enlightenment* describes three major tests during the eighteenth century of Newton's theory of universal gravitation.

The **controversy over the shape of the Earth**, which was the first of the three tests, was mentioned by Voltaire in the passage cited at the end of the last chapter. That famous French controversialist noted that "At Paris you imagine that the earth is shaped like a melon, or of an oblique figure; at London it has an oblate one."[7] Newton had claimed in his *Principia* that the

---

[5] Einstein and Infeld, *Evolution of Physics,* p. 33.

[6] Alexander Pope, *Minor Poems,* ed. by Norman Ault, completed by John Butt (Yale University Press, 1954), p. 317; see also Derek Gjertsen, *The Newton Handbook* (Routledge and Kegan Paul, 1986), p. 439.

[7] As given in Voltaire, *Letters on the English* in *French and English Philosophers*, vol. 38 of the Harvard Classics (P. F. Collier and Son, 1910), p. 109.

rotation of the Earth led to its bulging at the equator, in support of which he cited observations made in 1672 near the equator by Jean Richer, who found that a pendulum of some definite length would swing slightly more slowly at the equator than the same pendulum would swing in France. Newton explained this effect by attributing an equatorial bulge to the Earth, which would increase the distance of the pendulum bob from the Earth's center. This increase would (1) diminish the gravitational attraction on the pendulum bob, and (2) increase the centrifugal force on the bob resulting from the Earth's rotation. These two effects would combine to produce the observed change in the time of the swing of the pendulum.

Cartesians, however, believed that their vortex theory would lead to the Earth's being elongated toward its poles. In short, Newton attributed a tangerine shape to the Earth, whereas Cartesians favored a lemon shape. In 1718, two highly regarded French astronomers, Jacques Cassini and his father Jean Dominique Cassini, published observations supporting the Cartesian claim. Such conflicting observational reports led to a call for new measurements; in particular, scientists proposed that one expedition be sent to the equator, another toward the poles. Thus in the mid-1730s, one expedition headed for Ecuador, another toward the north polar region, the latter group being led by two prominent scientists, Pierre-Louis Moreau de Maupertuis and Alexis Clairaut, both of whom had become convinced Newtonians. The observations completed by both groups supported the Newtonian position, and carried credibility, which gave the Newtonians a major victory.[8]

The **controversy over the Moon's motions** reached its peak in 1747, when Clairaut reported to the French Academy of Sciences that Newton's gravitational theory would not fully account for the motions of the Moon. A number of factors made the prediction of the Moon's motions an especially significant issue. One was that the nearness of the Moon makes extremely accurate positional observations possible. Another was that having precise tables of the Moon's motions had important practical implications; such tables, it was believed, might offer a solution to the pressing nautical problem of determining longitude at sea. A third factor was that scientists recognized that calculating the Moon's motions, which are influenced by the gravitational attraction of *both* the Sun and the Earth (the so-called three body problem), was among the most complex mathematical issues of the day. Indeed, mathematicians realized that they could make their reputation by

---

[8]For more information on this and the other two tests, see Thomas Hankins, *Science and the Enlightenment* (Cambridge University Press, 1985), pp. 37–41.

developing a superior method for dealing with this problem. Consequently, three of the most skilled mathematicians, Clairaut, Jean Lerond d'Alembert, and Leonhard Euler, took up the challenge. All developed methods for treating the three body problem; moreover, d'Alembert and Euler came to agree with Clairaut that Newtonian methods were not adequate to this problem. In 1749, however, Clairaut reversed himself, concluding that Newtonian methods, rather than leading to incorrect conclusions, were fully satisfactory, d'Alembert and Euler agreeing.

Dramatic as these first two international controversies were, the **controversy over the time of the return of Halley's comet** was even more sensational in outcome. Decades earlier, Edmond Halley, based on an inspection of historical records indicating that a bright comet had appeared in the years 1456, 1531, 1607, and 1682, noted that a period of 75 to 76 years separated these apparitions and suggested that rather than these being distinct comets, they were one comet moving in an elongated ellipse and passing by the Sun at times separated by this period. Newton's views in his *Principia* were supportive of this, Newton having shown on theoretical grounds that comets must move in either parabolas, hyperbolas, or ellipses, the last figure leading to periodic returns. Moreover, Newtonian methods entailed the possibility of making the prediction of the return of a comet more precise by factoring in the effects of gravitational forces from planets on the comet's orbit. As the time for the next appearance of Halley's comet began to approach (1682 + 76 = 1758), Clairaut set about making such a precise calculation. He succeeded brilliantly in his prediction, an impressive testimony to his skill and also to the power of Newtonian methods.

The success of Newtonian mechanics in handling these challenges enhanced its reputation. Additional successes followed—for example, the achievement of Pierre Simon Laplace in working out his *Mécanique céleste*, a five volume treatment of celestial mechanics published between 1799 and 1825, and based on Newton's methods. Another very striking prediction occurred in 1846, when Urbain Jean Joseph Leverrier and John Couch Adams used Newtonian mechanics to predict the existence of the planet Neptune, which the observational astronomer Johann Galle located almost exactly in the predicted position.

Nonetheless, in the period between 1700 and 1900, it was becoming ever clearer that physics is a unified whole, that the various areas of physics—mechanics, heat, light, electricity, magnetism, etc.—form a unity and that any system of physics needs to be tested not just in a single area but against the whole range of physical phenomena. Although Newtonian mechanics had met with numerous successes, other aspects of the Newtonian system, e.g., its ideas regarding light, were coming to be challenged and in some cases were found to be faulty.

Heat Theory and the Concept of Energy

By the late eighteenth century, **Joseph Black** (1728–1799) and others had distinguished between the intensity of heat (temperature) and the quantity of heat in a body and had developed the notion of the specific heat or heat capacity of a body. This was largely done in the context of the **caloric theory of heat**, according to which heat consists of a weightless material substance, which functions as a conserved quantity in reactions involving heat. Put differently, there exists a material substance, caloric, such that it has the property that in reactions between substances, the quantity of caloric does not itself change.

In 1798, **Benjamin Thompson, Count Rumford** (1753–1814), raised problems for the caloric theory of heat by reporting experiments in cannon boring in which vast quantities of heat seemed to be created. This did not fit with the idea that heat is a conserved quantity or with the conception of heat as a weightless substance. Although Einstein and Infeld claim that Rumford's experiments were "crucial" and acted as a "death blow" to the caloric theory,[9] this is a serious overstatement, since the caloric theory was widely held until the 1840s. Only then did the **dynamical theory of heat** (the theory that heat consists of the motions of atoms) come to be accepted.

By 1850, the idea of **energy** had emerged and served to unify conceptually the various branches of physical science. **Gottfried Wilhelm Leibniz** (1646–1716) had earlier worked out the notion of what he called "*vis viva*" or "living force." He quantified this as mv^2 (m being mass and v being velocity) and suggested that it should be seen as the measure of the quantity of motion that God had placed in the universe. He did this in opposition to Descartes who had proposed mv as the measure of that total motion. This dispute extended well into the eighteenth century. Leibniz also stated that this force exists not only in a "living" form but also in a potential form. When, for example, an apple falls, its potential force is converted into its *vis viva*. These ideas eventually led to the notion, which emerged far more clearly in the nineteenth century, that there are two forms of mechanical energy, **potential and kinetic energy**. Early in the nineteenth century, the measure of kinetic energy was taken to be not mv^2, but rather $\frac{mv^2}{2}$.

In the 1840s, **Julius Robert Mayer** (1814–1878), **James Prescott Joule** (1818–1889), **Hermann von Helmholtz** (1821–1894), and others contributed to the introduction of the idea that there exists in the universe a conserved quantity, **energy**, that manifests itself in various forms, e.g., chemical, electrical, mechanical, thermal, etc. One such conversion is that

---

[9]Einstein and Infeld, *Evolution of Physics,* pp. 41–42.

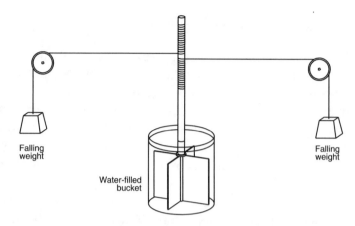

Diagram of Joule's Paddle Wheel Device for Measuring
the Mechanical Equivalent of Heat

between mechanical and thermal energy. Joule measured the **mechanical equivalent of heat**. He found that when a one pound body falls through 772 feet, it always generates exactly the same quantity of heat: specifically, the quantity of heat needed to raise the temperature of one pound of water by one degree Fahrenheit. Joule's device is shown in the accompanying diagram. As the weights fall, the paddle wheel turns through the bucket of water, producing a rise in temperature in the water. In this process, mechanical energy is converted into thermal energy. The introduction of the notion of energy played a major part in the eventual rejection of the caloric theory of heat and in its replacement by the kinetic or dynamical theory.

In the late nineteenth century, the dominant approach to physical science was the **mechanical view of nature**. Einstein and Infeld cite the following statement by Helmholtz as presenting the program of this point of view: "we discover the problem of physical material science to be to refer natural phenomena back to unchangeable attractive and repulsive forces whose intensity depends wholly upon distance. The solubility of this problem is the condition of the complete comprehensibility of nature."[10] This entails the notion that the proper explanation of a physical process is one reducing it to forces acting among unalterable bodies separated by certain distances. The **kinetic theory of gases**, the fundamental principles of which had been formulated by about 1870, was a prime example of the effectiveness of this approach. The kinetic theory explained the properties of gases by treating them as composed of atoms in rapid motion governed by mechanical laws.

---

[10] As quoted in Einstein and Infeld, *Evolution of Physics,* p. 54.

Electricity and Magnetism

By about 1820, scientists had worked out the mechanical view of nature quite effectively in regard to both electricity and magnetism. They showed, for example, that many electrical phenomena can be explained by treating electricity as though it were composed of particles charged either positively or negatively, which particles attract each other when of opposite charge but repel each other if of the same charge. The forces of attraction and repulsion are inversely proportional to the squares of the distances that separate the particles. This, however, entailed the postulation of two electrical fluids. These fluids were not conceptually related to mass, that is, they were weightless or "imponderable."

For a period around 1800, a key notion for physics was the doctrine of the imponderables. This was the doctrine that heat, electricity, magnetism, and also light consist of imponderable substances. Physical phenomena were to be explained in terms of the properties of these materials: caloric, the two electrical fluids, two magnetic fluids, and the particles making up light.

In 1820, **Hans Christian Oersted** (1777–1851) discovered a phenomenon that did not fit easily into the mechanical view of nature. He found that a current-carrying wire exerts a force on a magnet placed near it, which force is such as to make the magnet turn so that it lies at right angles to the current-carrying wire. **Henry Rowland** (1848–1901) further complicated this situation by showing that the forces on such a magnet depend on the velocity of the electrical flow. In particular, if a magnet is placed at the center of a rotating sphere, the exterior of which is covered with electrical charge, the magnet not only experiences a force, but the force becomes greater if the rate at which the sphere is rotating is increased. Nothing analogous to either of these effects was known in mechanics, where forces do not act at right angles and they do not depend on the velocities of the interacting objects.

Light: Particle or Pulse?

By 1650, a number of laws of light, e.g., the laws of reflection and refraction, had been established and various theories of light had begun to be developed. A major dispute on the nature of light broke out in the seventeenth century between Newton and such Cartesians as Christiaan Huygens. Put in a somewhat oversimplified form, Newton championed a particle theory of light according to which light rays consist of tiny particles that move at great speed. Huygens, on the other hand, held that light consists of pulses in the aether (which, according to Cartesians, pervades all space). Newton presented his view at first in two papers written during the 1670s and then in a far more fully developed form in his *Opticks* (1704). Huygens presented his views in his *Treatise on Light* (1690).

By the early eighteenth century, it had been shown that light moves with a finite but very large velocity. Most eighteenth-century optical theorists conceived light as consisting of particles. In the first three decades of the nineteenth century, however, **Thomas Young** (1773–1829) and **Augustin Fresnel** (1788–1827) successfully argued that light consists of transverse waves moving in an aether, which fills all space.

Among the most convincing pieces of evidences for this view was the ability of proponents of the wave theory of light to explain the apparent disappearance of light in certain experiments by the idea that light waves, when out of phase, can mutually destroy each other. The development of the wave theory of light involved the postulation of a weightless **aether**, and also the attribution to the aether of various hypothetical properties that some found difficult to accept. Einstein and Infeld view the **wave theory of light** as another victory for the mechanical view of nature.

Field Theory

In the period from around 1830 on, a new conceptual approach began to develop in physical science that differed in fundamental ways from the mechanical view of nature. This approach, which came largely from researches in electricity, magnetism, and optics, centered on the notion of a **field**. Emphasis was shifted away from, for example, the particles of electricity and placed on the structure of the field surrounding the particles. **Michael Faraday** (1791–1867) was an important contributor in this

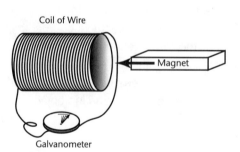

Coil of Wire

Magnet

Galvanometer

area. He discovered that a magnet moving in the vicinity of a closed wire loop produces a current in the wire. He interpreted this in terms of electrical and magnetic lines of force, which lines represent the structure of the field. **James Clerk Maxwell** (1831–1879) developed Faraday's ideas in a detailed mathematical form and presented theoretical arguments for interpreting light as an **electromagnetic wave**. **Heinrich Hertz** (1857–1894) provided experimental verification that light is an electromagnetic wave. Tensions between the mechanical view of nature and the field theory approach consequently arose. These stimulated Einstein to a reformulation of much of classical physics.

Part II: A Development of Particular Importance: The Michelson-Morley Experiment and Early Attempts to Deal with Its Implications[11]

Some Mathematical Background for Comprehending the Michelson-Morley Experiment and Also Einstein's Special Theory of Relativity

As a preparation for understanding the famous Michelson-Morley experiment and the efforts to explain its results theoretically, it is helpful to discuss some fairly elementary mathematics. In particular, we will attempt to solve a problem that sets the stage for the Michelson-Morley experiment and also discuss what Einstein called the Galilean transformation equations. In doing this the clearest way is to use elementary vectorial methods for the solution of this problem. A second very important benefit of understanding these discussions is that they pave the way for understanding Einstein's special theory of relativity.

Brief Explanation of Vectors

Newton's first corollary to his laws of motion states: "*A body [urged] by forces joined together, describes the diagonal of a parallelogram in the same time in which it describes the sides separately.*" The frequency with which physicists and mathematicians need to deal with the combination or composition of forces, velocities, and other directed entities led to the development during the nineteenth century of a form of mathematics specifically designed to facilitate working with such magnitudes: this area is called vector analysis. Vectors have proven very useful in efforts to represent and analyze various phenomena. Vectors are mathematical quantities that have not only a certain magnitude but also a direction associated with them. Among the entities that vectors very effectively represent are velocities and forces. Thus one can describe a velocity as having the magnitude 20 mph and a direction due north.

To take an example, imagine an object given a velocity of 10 mph in an eastward direction and also a velocity of 10 mph in a northerly direction. Those velocities, if added as vectors, will give the resultant velocity. In the diagram (next page), these two velocities are represented as arrows

---

[11] The materials in this section and in the following chapter are based in part on the exposition in Gerald Holton, *Introduction to Concepts and Theories in Physical Science,* revised by Stephen G. Brush, 2nd ed. (Addison-Wesley, 1973), Ch. 30.

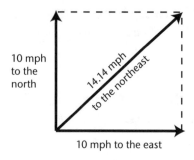

as is their sum, which is the diagonal of the parallelogram of which the two vectors form the sides. This is true in general: to find the resulting velocity of a body subjected to two differently directed velocities, add those two velocities as vectors. The resulting vector will represent the new velocity.

Problem to Be Solved

Suppose your home is located on a river and that you urgently need to make a purchase at a store that sells the item needed. In fact, there are two stores, each one mile distant. One store (call it the Cross-Stream Cafe) is directly across the river; the other (call it the Down-Stream Diner) is located 1 mile downstream. Suppose also that the river flows at the rate of 3 miles per hour and that you are able to row through the water at 5 miles per hour. Will it take less time to row across the river and back or to row downstream (with the current) and back (against the current)?

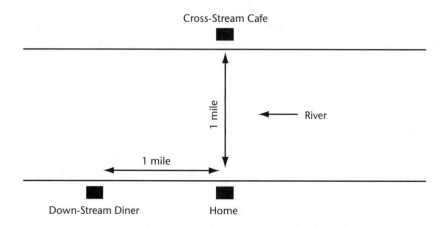

First Solution   (1) First we will calculate the time it takes to row to the Down-Stream Diner and to return. If you row downstream, you will move at the rate of 3 mph + 5 mph = 8 mph in relation to the dock from which you departed. On the return trip, you will move at the rate 5 mph − 3 mph = 2 mph. Hence the total time will be:

$$T_1 = \frac{1 \text{ mile}}{8 \text{ mph}} + \frac{1 \text{ mile}}{2 \text{ mph}} = \frac{5}{8} \text{ hours}$$

(2) Now we will calculate the time it takes to row to the Cross-Stream Cafe and return. Were you to attempt to row directly across the river pointing the boat at the immediately opposite shore, you would not in fact move in a perpendicular line. You would end up moving in an oblique path, because for every 5 yards you would row across the river, you would be carried by the river 3 yards downstream. Consequently you must point the bow of your boat to some extent upstream. We can diagram this as follows:

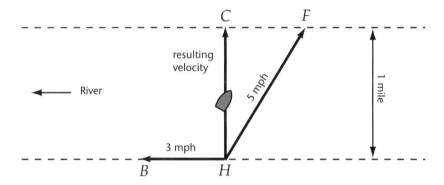

If you point your boat in the direction HF, where F is positioned so that $CF = BH$, this will give your boat an upstream velocity equal to HB, the downstream velocity of the river. It is a fundamental theorem of mechanics that velocities can be added like vectors. Hence the vector HC, which is the diagonal of the parallelogram $HBCF$, can be seen as representing the sum of vectors HF and HB. Because $HB = CF$ and because HC is perpendicular to HB and to CF, we have, by the Pythagorean theorem,

$$HF^2 = CF^2 + HC^2$$

or $(5 \text{ mph})^2 = (3 \text{ mph})^2 + V_{HC}{}^2$, where V_{HC} is the velocity in the direction HC. Hence $V_{HC}{}^2 = 25 - 9 = 16$, and $V_{HC} = 4$ mph.

We see that you will move across the river at the rate of 4 mph, provided that you row as indicated. By a comparable analysis, it is easy to see that the return trip should also be at the rate of 4 mph. By the formula $VT = D$ (velocity times time equals distance), we have 4 mph $\cdot$ $(T_2) = 2$ miles. Consequently, your total time to cross the river and return, will be

$$T_2 = \frac{1}{2} \text{ hour.}$$

Note that this is significantly less than the time taken by rowing down and then up stream. Thus it will be more efficient if you row across the river and back, rather than row downstream and return.

We can also analyze this situation from the perspective of the river, so to speak. This we will do in the next section.

Second Solution   Having first analyzed this problem from the point of view of the velocities, we can now consider it in terms of the distances involved, concerning ourselves primarily with the cross-river motion. Whereas in the earlier analysis, the fixed point of reference was the place on the shore from which the boat started, in this analysis, the point to which motion will be referred is the moving river or, more precisely, a drop of water moving along with that river.

Let V_R be the river velocity, 3 mph. Let the effective distance that the boat must travel in crossing the river and returning be d. This will also be the path of the boat as seen from a drop of water moving with the same velocity as the river; thus from the point of view of the drop of water, the boat will seem to be moving along the path indicated by the arrows. Let T represent the total time of rowing down and back, or across and back. The problem can be conceptualized according to the following diagram:

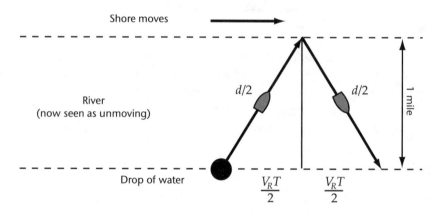

(1) **Down-Stream Case:** From the point of view of the drop of water, the boat will move away from it (moving downstream) at the rate of 5 mph. The boat will continue to move downstream until it gets to the Down-Stream Diner one mile downstream. The time necessary for it to reach that point will be $\frac{1}{5+3} = \frac{1}{8}$ hrs., the reason being that the motion of the boat and of the river will complement each other, giving the boat effectively a velocity of 8 mph in relation to your dock. The boat will then reverse direction and will be seen from the point of view of the drop to move at 5 mph. It will move at that rate until it reaches your dock, which will take $\frac{1}{5-3} = \frac{1}{2}$ hrs., the reason being that

the river's motion will impede the boat's efforts to move upstream, making its velocity $(5\,\text{mph} - 3\,\text{mph})$ or 2 mph. To put it differently, your dock is constantly moving away from the drop, making it necessary that your boat's motion compensate for the apparent motion of the dock. Thus the total time of travel will be

$$T_1 = \frac{5}{8} \text{ hrs.}$$

(2) **Cross-Stream Case:** From the point of view of a drop of water located in the river, the boat that crosses the river to the Cross-Stream Cafe must move as indicated over the distance $\frac{d}{2} + \frac{d}{2} = d$. The downstream distance due to the motion of the river is $V_R T$. In other words, from the point of view of the drop of water, the boat at the end of its excursion when it has returned to your dock will be a distance $V_R T$ farther upstream than the drop of water. By the Pythagorean theorem, we have

$$\left(\frac{d}{2}\right)^2 = 1^2 + \left(\frac{V_R T}{2}\right)^2.$$

Hence $\dfrac{d^2}{4} = 1 + \dfrac{V_R{}^2 T^2}{4}$. Because $V_R = 3$, we have $\dfrac{d^2}{4} = 1 + \dfrac{9T^2}{4}$. Multiplying this equation by 4 gives $d^2 = 4 + 9T^2$. But we know that

$$d = T \cdot 5 \text{ mph,}$$

which by substitution gives us

$$(5T)^2 = 4 + 9T^2,$$

that is,

$$25T^2 - 9T^2 = 4.$$

This is equivalent to

$$16T^2 = 4 \quad \text{or} \quad T^2 = \frac{1}{4},$$

which gives us

$$T_2 = \frac{1}{2} \text{ hour,}$$

the same result that was found previously.

Understanding both these solutions to the problem will significantly help in following the analysis of the Michelson-Morley experiment presented in the next section. One particular feature of this problem situation deserves to be highlighted: this situation provides a method of detecting whether you live on a lake or a river. If the travel time for a one mile journey across the body of water and back is equal to the travel time for a one mile journey parallel to the shore and back, then one lives on a lake. If the times differ, then one lives on a river. This may seem obvious, but in what follows, we shall encounter a situation where application of this method produced a shocking result.

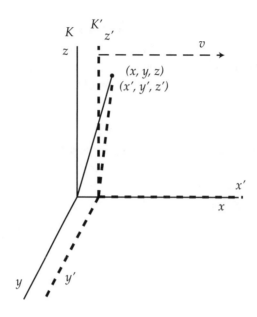

The Galilean Transformation Equations

Einstein discussed what he called the "Galilean transformation equations." It should not be assumed that Galileo explicitly laid out these equations. Rather, they were implicit in his work and in essentially all books on mechanics written in the centuries after Galileo. What is significant from our point of view is that Einstein proposed a radical revision of these equations. Consequently, comprehension of these equations is helpful for understanding the Michelson-Morley experiment and its early interpretation and is vital for understanding the special theory of relativity.

The Galilean transformation equations provide a method of describing the position, motion, etc., of an object moving in a frame of reference that is itself moving at a constant velocity; for example, we can use them to describe the motion of a woman walking north on a train that is itself moving northward. Let the velocity of the woman relative to the train be 3 miles per hour and suppose the train is moving at 60 mph. Relative to the train station, she will be moving at $60 + 3$ or 63 mph. To express this situation in general terms, suppose we have two frames of reference or coordinate systems K and K'. K' is moving with velocity v in the x direction relative to the K frame (see the diagram above).

First of all, let us state the equations that permit us to fix the position of a point in the moving K' frame with reference to the stationary K frame after the K' frame has moved away from the origin of the K frame during some time period t. The coordinates of a point in the K' frame will be:

$$x' = x - vt$$
$$y' = y$$
$$z' = z$$
$$t' = t$$

Comment: As an illustration of the first equation, consider a K' frame attached to the stern of a boat moving away from a dock at 20 mph. The boat has been out of port (the K frame) for 2 hours. The coordinate of the stern of the boat in the K' frame after time t will be: $0 = x - 20 \cdot 2$. This tells us that although the position of the stern will still be at the origin in the K' frame, the stern seen from the K frame will be 40 miles distant. No change occurs in the y and z coordinates; moreover, observers in both frames will report identical time determinations: if the boat left at noon, it will now be 2 p.m. for both the pilot and a person on the dock.

We can also state equations that give the coordinates in the K frame of any point in the K' frame of reference.

$$x = x' + vt'$$
$$y = y'$$
$$z = z'$$
$$t = t'$$

Problem

Imagine a rod of length L lying along the x axis in the K frame. The length of the rod will be $L = x_2 - x_1$, where x_1 and x_2 are simply the coordinates of its end points. Suppose an observer in the K' frame wishes to measure the length L' of the rod relative to his or her moving frame of reference; this will require determining x_1' and x_2', the coordinates of the rod's end points as seen in the observer's frame.

Solution: We have $L' = x_2' - x_1'$. Employing the substitutions provided by our first set of transformation equations, we get $L' = (x_2 - vt) - (x_1 - vt)$ $= x_2 - x_1$. Consequently, we have the result that $L' = L$, i.e., the length of the rod will appear to be the same to observers in either the fixed or the moving frame.

Comment: All this seems obvious enough. Why, after all, should the length of a rod seem to change when it is set in motion? The relevance of these equations is that in Einstein's special theory of relativity, the equations must be revised. In fact, Einstein maintains that the measured length of a rod *will* change when the rod is set into motion with respect to the measuring observer!

Questions Concerning the Aether

As the wave theory of light developed, questions arose concerning the relation of the waves of light to the aether, that weightless and invisible material that, spread throughout the universe, was conceived to be the medium in which light waves move. We can get some idea of what is involved from considering the propagation of sound. Suppose that on a windless day we send a sound signal with a velocity V_s over some distance D. Then, because the sound is moving at velocity V_s, the time of travel will be $\frac{D}{V_s}$. Suppose now that a wind is blowing in the direction from the source of the sound to the receiver, the wind having a velocity V_w. Then, because the sound is moving through the air at a velocity V_s and the air itself is moving at a velocity V_w, the time of transit will be $\frac{D}{V_s+V_w}$. In this case it will take less time for the sound to move from the source to the receiver. Or suppose that the receiver is moving toward the sound source at a velocity V_r. Then the time will again be less; it will equal $\frac{D}{V_s+V_r}$.

With this as background, we can examine the options faced by nineteenth-century physicists when they thought about the motion of the Earth relative to the aether. Three possibilities present themselves.

1.   The **Earth moves freely through the aether**. Thomas Young, one of the founders of the wave theory of light, favored this possibility. Early in the nineteenth century, Young stated that "I am disposed to believe that the luminiferous aether pervades the substance of all material bodies, with little or no resistance, as freely, perhaps, as the wind passes through a grove of trees."[12]

2.   Another idea that was considered is that somehow **the Earth drags the aether** along with it, just as air in a train car is dragged along with the car.

3.   As a final possibility, it is conceivable that the **aether is partially dragged** along with the Earth.

It became of much interest to decide among these three possibilities. The first was particularly attractive because, were it to be the case, the aether would provide an absolute frame of reference. According to the first alternative, light from a star should exhibit different velocities depending on whether we are moving toward or away from it, since the Earth's motion

---

[12] As quoted in Loyd Swenson, *The Ethereal Aether: A History of the Michelson-Morley-Miller Aether-Drift Experiments, 1880–1930* (University of Texas Press, 1972), p. 18.

would become compounded with the velocity of light from the star. But during the nineteenth century, the French astronomer François Arago (1796–1853) and others showed empirically that this is not the case. Given the improbability of the second hypothesis (an aether fixed to the Earth), interest was strongest in the third hypothesis—that of a partially dragged aether.   The French physicist Augustin Fresnel in the 1820s developed this hypothesis of an aether drag in a quantitative way that seemed to work within the limits of observation.

In the second half of the nineteenth century, James Clark Maxwell and others proposed more exact experiments, and in the 1880s Albert Michelson attempted to provide an experimental test. Although an American, Michelson first attempted the experiment in Potsdam in 1881. He then tried it again in 1887 working in conjunction with Edward W. Morley at Case University in Cleveland.

The Michelson-Morley Experiment

Note: In coming to an understanding of the following presentation of the Michelson-Morley experiment, you will be helped by reviewing the boat problem presented earlier in this chapter.

The 1887 experiment carried out by **A. A. Michelson** and **E. W. Morley** took essentially the following form: they placed two mirrors M_1 and M_2 at equal and perpendicular distances L_1 and L_2 from a partially silvered mirror. Light from a source S, partially transmitted and partially reflected by this mirror, proceeded onward to strike mirrors M_1 and M_2.

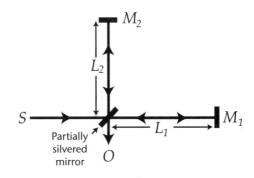

From them it was reflected back to the first mirror and projected onto a viewing screen at O.

If the round-trip travel times of the rays along paths L_1 and L_2 are equal, then the two returning light waves will recombine and be in phase at every point on the viewing screen, producing uniform illumination of the screen. Any inequality between the times, however, will cause the returning waves to be out of phase at some points on the viewing screen, but in phase at other points. This produces an **interference pattern**, consisting of alternating light and dark fringes, on the screen. In an ideally adjusted instrument, the number and spacing of these fringes are related to the magnitude of the time difference and can therefore serve, in principle, as a measure of that difference.

Let us assume the Earth's motion is such that the instrument is moving to the right at some velocity v in relation to the aether. Another way of viewing this is that the aether is flowing to the left past the instrument at the velocity v (recall the second boat problem discussed earlier). Then the mirror M_1 will be moving away from the point where the source S at some instant emitted a ray. This will make the travel times for the first part of this trip longer and the second part (return) shorter. Let us calculate this amount, using c to represent the speed of light. Let T_1 be the time taken by the ray in moving in the horizontal direction in the figure. Then

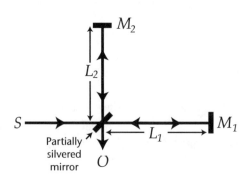

$$T_1 = \frac{L_1}{c-v} + \frac{L_1}{c+v}.$$

We can simplify this equation by seeking a common denominator for its two parts.

$$T_1 = \frac{L_1}{c-v} + \frac{L_1}{c+v} = \frac{L_1}{c-v} \cdot \frac{c+v}{c+v} + \frac{L_1}{c+v} \cdot \frac{c-v}{c-v}$$

$$= L_1 \left(\frac{c+v}{c^2 - v^2} + \frac{c-v}{c^2 - v^2} \right) = L_1 \left(\frac{c+v+c-v}{c^2 - v^2} \right) = \frac{2L_1 c}{c^2 - v^2}$$

$$= \frac{2cL_1}{c^2 \left(1 - \frac{v^2}{c^2}\right)} = \frac{2L_1}{c} \cdot \frac{1}{1 - \frac{v^2}{c^2}}.$$

We shall now calculate the time, T_2, taken by the ray moving in the vertical direction. As suggested by the earlier analysis of boat travel, the path of the ray, if considered from the point of view of a fixed point in space, will be as indicated in the next diagram. (Compare it with the diagram for the boat problem on page 258.) You can see that if the ray is to strike mirror M_2, it must proceed at a small angle to the vertical.

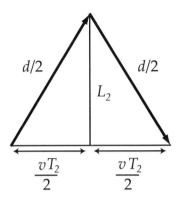

We know that $T_2 = \frac{d}{c}$ where d is the distance the ray travels. By the Pythagorean theorem, we have: $(\frac{d}{2})^2 = (\frac{vT_2}{2})^2 + L_2^2$. This implies that $d^2 = v^2T_2^2 + 4L_2^2$, which in turn implies that

$$d = \sqrt{v^2T_2^2 + 4L_2^2}.$$

We now substitute this value into the equation $T_2 = \frac{d}{c}$.

This gives $T_2 = \frac{d}{c} = \frac{\sqrt{v^2T_2^2 + 4L_2^2}}{c}$.

If we square both the leftmost and rightmost sides of this equation, then multiply both sides by c^2, and then subtract $v^2T_2^2$ from both sides, we obtain

$$c^2T_2^2 - v^2T_2^2 = 4L_2^2.$$

Factoring the left side of this gives:

$$T_2^2(c^2 - v^2) = 4L_2^2.$$

Thus

$$T_2^2 = \frac{4L_2^2}{c^2 - v^2}.$$

Taking the square root of both sides gives

$$T_2 = \frac{2L_2}{\sqrt{c^2 - v^2}} = \frac{2L_2}{c} \cdot \frac{1}{\sqrt{1 - \frac{v^2}{c^2}}}.$$

We wish to compare T_1 and T_2. If we find that they are equal, this suggests that in relation to the apparatus, the aether forms in effect a lake, but if they turn out to be unequal, it suggests that the aether corresponds to a river through which we move.

Because the apparatus was constructed so that $L_1 = L_2$, we can replace them by a generalized length L. Then we have:

$$T_1 - T_2 = \frac{2L_1}{c} \cdot \frac{1}{1 - \frac{v^2}{c^2}} - \frac{2L_2}{c} \cdot \frac{1}{\sqrt{1 - \frac{v^2}{c^2}}} = \frac{2L}{c} \cdot \left(\frac{1}{1 - \frac{v^2}{c^2}} - \frac{1}{\sqrt{1 - \frac{v^2}{c^2}}} \right)$$

Consequently, if we are moving through an aether with velocity v, a small but definite difference should exist. The difference will indeed be small because in the fraction $\frac{v^2}{c^2}$, c is very large (the speed of light is 186,000 miles per second). Nonetheless, the difference should be detectable by the appearance of interference fringes. Michelson and Morley, however, found no effect, even when the apparatus was rotated through $90°$. How was this startling result to be explained? Three possibilities occurred to scientists of the period, although none seemed fully satisfactory.

1. The aether surrounding the Earth is completely dragged with the Earth. This assumption created problems because it predicted observations that would contradict the well known phenomenon of stellar aberration, which is a detectible shifting of the observed position of stars. In 1728, the British astronomer James Bradley had not only discovered this phenomenon but also showed that it is due to the motion of the Earth around the Sun.

2. The motion of the source should be added. It may have occurred to you that if the velocity of the source is added to the velocity of light, that would account for the Michelson-Morley null result. But this will not work either, because light is a wave in the aether, and it is the nature of waves to move with their characteristic velocity in relation to the medium in which they move, as the speed of sound is related to the air in which the sound travels.

3. *The Lorentz-FitzGerald Contraction.* This third hypothesis (which will be discussed shortly) was quite radical and was, moreover, criticized as *ad hoc,* meaning that it was constructed solely to account for this result.

Exercise

Check the correctness of the equations formulated in this analysis by applying them to the earlier problem of rowing on a river. It should be evident that the two paths of the boat (horizontal and vertical) correspond to the two paths of the ray of light and that the dock corresponds to the first mirror.

The Lorentz-FitzGerald Contraction

In 1889, the Irish physicist **George FitzGerald**, and later the Dutch physicist **Hendrik Lorentz**, suggested that the null result could be explained by assuming that when bodies move through the aether with the velocity v, they contract by a factor that is equal to $\sqrt{1 - \frac{v^2}{c^2}}$. This contraction, called the **Lorentz-FitzGerald contraction**, serves to account for the phenomenon because it makes the length of the horizontal path $= L_1\sqrt{1 - \frac{v^2}{c^2}}$.

Consequently, if we multiply the term corresponding to T_1 by $\sqrt{1 - \frac{v^2}{c^2}}$, we get:

$$T_1 - T_2 = \frac{2L_1}{c} \cdot \frac{\sqrt{1 - \frac{v^2}{c^2}}}{1 - \frac{v^2}{c^2}} - \frac{2L_2}{c} \cdot \frac{1}{\sqrt{1 - \frac{v^2}{c^2}}} = 0,$$

which fits with the null result in the Michelson-Morley experiment. But of course the Lorentz-FitzGerald contraction hypothesis does not explain *why* this contraction should occur. The most famous response to that question came from the pen of the young physicist central to the next chapter: Albert Einstein.

Albert Einstein

Chapter 6

Einstein and Relativity Theory

Introduction

In 1905, Albert Einstein published his special theory of relativity. The significance of this event for the history of mechanics can be compared only with Galileo's publication in 1638 of his *Two New Sciences* and Newton's publication in 1687 of his *Principia,* which is to say that it was an epoch-making event. In fact, Einstein's special theory of relativity was vitally important not only for mechanics but for other areas of physics as well.

The drama involved in this famous event should not be missed. When Einstein published his special theory of relativity, he was very young and was working as a patent inspector in the Swiss Patent Office in Bern, Switzerland. He had failed to secure a teaching position, and completed his doctorate only in 1905 (a previous thesis had been turned down in 1901). Improbable as it may seem, Einstein's special theory of relativity paper was one of four papers that he published in 1905, any one of which would have assured his fame. It was indeed an *annus mirabilis,* which is often compared to Newton's miraculous year.

The drama is further enhanced by the fact that Einstein successfully challenged Newtonian mechanics, which many saw as the very embodiment of scientific certainty. Perhaps most striking is the fact that Einstein took the problem of the relativity of motion, which had beset physical thought for millennia, and in a sense made this difficulty the cornerstone of his theory, showing that if properly understood, it would reveal a great deal about the physical world.

The monumental scientific and philosophical significance of Einstein's special theory of relativity often leads people to believe that it is extremely difficult to comprehend. The following pages embody my conviction that this is not the case, that persons who carefully read them will be able to understand the special theory of relativity in a reasonably deep and satisfying manner. One reason for this is that previous chapters have set the stage for this presentation of relativity theory. Another reason is that the mathematical treatment of the special theory of relativity requires only ordinary algebra and geometry. The mathematics is not easy, but is quite straightforward. The presentation that follows is based on two of Einstein's books: *Relativity: The Special and the General Theory* and *Evolution of Physics,* the latter having been jointly authored with Leopold Infeld. It also draws on the exposition in Gerald Holton, *Introduction to Concepts and Theories in Physical Science,* revised by Stephen G. Brush. This chapter also includes a brief and elementary summary of the general theory of relativity of 1916. Although comprehension of the general theory of relativity is a difficult task, some of its fundamental ideas are nevertheless within reach.

Chronology of the Life of Albert Einstein[1]

1879	Birth on March 14 in Ulm in southern Germany as the first child of Hermann Einstein, a chemical engineer, and Pauline Koch Einstein. His parents, both of whom were of Jewish descent, had married in 1876.
1880	The Einstein family moves to Munich.
1881	Birth of sister, Maja, who would later receive a doctorate in literature. In 1939, after Einstein takes up residence at Princeton, she comes to live with him until her death in 1951.
1884	Expresses amazement when shown a pocket compass; he later writes: "A wonder of such nature I experienced as a child of 4 or 5 years, when my father showed me a compass. That this needle behaved in such a determined way [always pointed to the north] did not at all fit into the nature of events, which could find a place in the unconscious world of concepts (effect connected with direct 'touch'). I can still remember—or at least believe I can remember—that this experience made a deep and lasting impression on me. Something deeply hidden had to be behind things."[2]
1885	Begins violin lessons.
1885–9	Attends a Catholic elementary school in Munich.

---

[1]This chronology is based on a number of sources, including the detailed chronology of Einstein's life in Abraham Pais, *"Subtle is the Lord...": The Science and the Life of Albert Einstein* (Oxford University Press, 1982).

[2]Albert Einstein, "Autobiography" in *Albert Einstein: Philosopher-Scientist,* ed. by Paul Arthur Schilpp, vol. 1 (Harper and Row, 1959), p. 9.

1889–94   Attends Luitpold Gymnasium, which would later be renamed the Albert Einstein Gymnasium.

1891   Albert's second great amazement: "At the age of 12 I experienced a second wonder of a totally different nature: in a little book dealing with Euclidean plane geometry, which came into my hands. . . . Here were assertions, as for example the intersections of the three altitudes of a triangle in one point, which—though by no means evident—could nevertheless be proved with such certainty that any doubt appeared to be out of the question. This lucidity and certainty made an indescribable impression upon me." [3]

1892   Reads various writings of the philosopher Immanuel Kant. Also develops an interest in the calculus.

1894   After his father's business fails, parents move to Milan, Italy. Albert joins them six months later, after leaving Luitpold Gymnasium without graduating.

1895–6   Fails entrance exam for Zurich Polytechnic or E. T. H. (Eidgenössische Technische Hochschule); enters the cantonal school in Aarau, Switzerland, from which he receives a diploma.

1896   Renounces his German citizenship, remaining stateless for five years; passes entrance exam and enters E. T. H., where he attends lectures by Hermann Minkowski. Meets Mileva Maric (1875–1948), another physics student, whom he would marry in 1903.

1900   Receives diploma from E. T. H.; unsuccessfully attempts to secure a position as an assistant there.

1901   Becomes a Swiss citizen (and remains a Swiss citizen throughout his life, even after becoming a U.S. citizen in 1940); completes his first scientific paper: "Consequences of Capillary Phenomena." Submits doctoral thesis for the University of Zurich, but thesis is turned down.

1901–2   Takes various temporary teaching positions at a number of locations.

1902   Birth of a daughter, Lieserl, to Einstein and Mileva Maric. No known records reveal the future life of this daughter. Accepts probationary position as a technical expert third class in the **Patent Office** in Bern, Switzerland. Einstein's father dies.

1903   Marries Mileva Maric on Jan. 6.

1904   Receives permanent appointment at the Patent Office; birth of first son, Hans Albert (1904–1973), who in 1936 would receive

---

[3] Einstein, "Autobiography," p. 9.

a doctorate from E. T. H. and would serve as professor of hydraulic engineering at the University of California (Berkeley) from 1947 to 1971.

1905   University of Zurich accepts his doctoral thesis ("On a New Determination of Molecular Dimensions") and awards his doctorate (1906). The year 1905 is also Einstein's *annus mirabilis*. He publishes four famous papers in the *Annalen der Physik*:

(1) "On a Heuristic Interpretation of the Generation and Transformation of Light," in which he shows Max Planck's new quantum theory can be used to explain the photo-electric effect.

(2) "The Motions of Particles Suspended in Liquids Deduced from the Kinetic Theory of Heat," in which he gives a theoretical account of Brownian motion.

(3) "On the Electrodynamics of Moving Bodies," in which he presents his Special Theory of Relativity.

(4) "Does the Inertia of a Body Depend on Its Energy Content?" in which he derives his famous equation $E = mc^2$.

1907   Formulates the **principle of the equivalence of uniformly accelerated systems**, a key idea in the general theory of relativity. Extends this principle to electromagnetic phenomena. Concludes that this principle (as extended) should entail the result that rays of light passing massive bodies should be bent, but believes the effect to be too small to be detectable.

1909   Leaves Patent Office for position at University of Zurich. The University of Geneva awards him his first honorary doctorate. He would receive at least eighteen more in subsequent years.

1910   Birth of second son, Eduard, who would eventually develop schizophrenia and die in a mental hospital in 1965.

1911   Concludes that his idea of the bending of light rays passing massive objects should be testable; privately predicts that light rays passing the Sun should be bent by 0.83 seconds of arc (an amount about half the accepted number).

1911   First Solvay Conference. Walther Nernst, with funding from the industrialist Ernest Solvay, organizes a gathering of the leading physicists to discuss the latest developments in science. Held in Brussels with Einstein among the select group invited.

1911–2   Takes position as physics professor at the University of Prague.

1912   Becomes professor at E. T. H.

1913   Publishes preliminary paper with Marcel Grossmann on **general theory of relativity**.

1913   Second Solvay Conference.

1914 Accepts position in **Berlin** as professor in the Prussian Academy of Sciences and Director of the Kaiser Wilhelm Institute of Physics. Einstein and his wife separate; she returns to Zurich with their two sons.

1915 Finds that general theory of relativity will account for the excess in the **advance of the perihelion of Mercury**; predicts that bending of light rays will be twice what he had previously thought.

1916 Publishes his **first major paper on general theory of relativity**.

1917 Publishes paper on cosmological implications of general theory of relativity.

1919 Feb. 14: Einstein and Mileva are divorced. May 29: **total eclipse of the Sun**, allowing **measurement of bending of light rays** passing near the Sun. June 2: marries his cousin Elsa, who has two daughters by a previous marriage. Nov. 6: joint meeting of Royal Society and Royal Astronomical Society at which it is announced that the measured result confirms Einstein's prediction. *Times* of London headlines: "Revolution in Science," "New Theory of the Universe," and "Newtonian Ideas Overthrown." Einstein becomes world famous.

1920 Einstein's mother dies. He meets Niels Bohr, a prominent Danish physicist. Disturbance, which may have been due to anti-Semitism, occurs at lecture by Einstein at University of Berlin. Johannes Stark and Philipp Lenard, two Nobel Prize winning German physicists, criticize Einstein's physical ideas, referring to them as "Jewish physics." Einstein's pacifism is also subjected to criticism.

1921 For the first time visits U.S., coming in an effort to raise funds for the planned Hebrew University in Jerusalem. Receives an extremely warm welcome from the public. Lectures at Princeton University.

1922 Receives the **Nobel Prize** in physics for his research on the theory of the photoelectric effect. Visits and gives lectures in Japan.

1923 Returning from the Orient, stops to visit Palestine and meets with various leaders of the Zionist movement.

1930–1 Makes trips to the U.S., spending time especially at the California Institute of Technology and at Princeton.

1932 Accepts professorship at the **Institute for Advanced Study** in Princeton, New Jersey, the original plan having been that he would divide his time between Princeton and Berlin.

1933	Jan. 30: **Nazis** come to power. Einstein resigns from Prussian Academy. Takes up temporary residence in Belgium. Leaves Europe on Oct. 7, never to return. Takes up residence in Princeton.
1934	Publishes *The World As I See It,* a collection of his essays.
1936	Death of Elsa Einstein, his second wife.
1938	In conjunction with Leopold Infeld, publishes *The Evolution of Physics.*
1939	At urging of Leo Szilard, sends letter to President Roosevelt, discussing the **military implications of atomic energy**. This letter eventually leads to the **Manhattan Project**, which results in the U.S. constructing atomic bombs.
1940	Becomes U.S. citizen.
1944	Prepares a handwritten copy of his 1905 paper on the special theory of relativity, which he donates to raise money for the war effort; when auctioned, it sells for six million dollars. It is now preserved at the Library of Congress.
1946	Assumes chairmanship of Emergency Committee of Atomic Scientists.
1949	Publishes his *Autobiography.*
1950	Publishes *Out of My Later Years,* a collection of his essays.
1952	Is offered presidency of Israel; declines.
1955	April 18: Dies in Princeton, New Jersey.

The Special Theory of Relativity

Introduction

In 1905, Albert Einstein published a paper with the seemingly innocuous title "Zur Elektrodynamik bewegter Körper" ("On the Electrodynamics of Moving Bodies"), in which he presented his special theory of relativity in a manner accessible only to persons with substantial background in physics. In a subsequent publication, *Relativity: The Special and the General Theory,* Einstein formulated his theory in a decidedly more accessible manner. The approach taken by Einstein in that book (and followed here) is to show that the Lorentz-FitzGerald contraction as well as much else can be deduced from two postulates that serve as the foundation for his theory. It is important to recognize that Einstein did *not* develop his theory from the Michelson-Morley experiment (discussed in Chapter 5); in fact, he later stated that it played little role in his thought.[4] Despite this, many writers present Einstein's

---

[4]Gerald Holton, "Einstein, Michelson, and the Crucial Experiment," *Isis, 60* (1969), 122–197.

theory as though it had been based on that experiment because it provides a direct and concrete way of grasping the fundamental ideas.

Einstein came to his theory from a number of different directions. Among these were the philosophical critiques of science formulated by **David Hume** (1711–1776) and by **Ernst Mach** (1838–1916). For example, in his *Autobiography,* Einstein stated that Mach had led him to question some fundamental aspects of classical physics: "It was Ernst Mach who, in his *History of Mechanics,* shook this dogmatic faith; this book exercised a profound influence upon me. . . while I was a student."[5] Mach led Einstein to question such matters as absolute space and time.

Tensions between Newtonian Mechanics and Maxwellian Electromagnetic Theory

Einstein was also led to his theory by tensions between Newtonian mechanics and the theory of the electromagnetic field as developed by Faraday and Maxwell. In Einstein's *Evolution of Physics,* he discussed these tensions in some detail. Moreover, this point emerges clearly from the first two paragraphs of Einstein's 1905 paper. There he wrote:

> It is known that Maxwell's electrodynamics—as usually understood at the present time—when applied to moving bodies, leads to asymmetries which do not appear to be inherent in the phenomena. Take, for example, the reciprocal electrodynamic action of a magnet and a conductor. The observable phenomenon here depends only on the relative motion of the conductor and the magnet, whereas the customary view draws a sharp distinction between the two cases in which either the one or the other of these bodies is in motion. For if the magnet is in motion and the conductor at rest, there arises in the neighbourhood of the magnet an electric field with a certain definite energy, producing a current at the places where parts of the conductor are situated. But if the magnet is stationary and the conductor in motion, no electric field arises in the neighbourhood of the magnet. In the conductor, however, we find an electromotive force, to which in itself there is no corresponding energy, but which gives rise—assuming equality of relative motion in the two cases discussed—to electric currents of the same path and intensity as those produced by the electric forces in the former case.
>
> Examples of this sort, together with the unsuccessful attempts to discover any motion of the earth relatively to the "light medium," suggest that the phenomena of electrodynamics as well as of mechanics possess no properties corresponding to the idea of absolute rest. They suggest rather that, as has already been shown to the first order of

---

[5] Einstein, "Autobiography," p. 21.

small quantities, the same laws of electrodynamics and optics will be valid for all frames of reference for which the equations of mechanics hold good. We will raise this conjecture (the purport of which will hereafter be called the "Principle of Relativity") to the status of a postulate, and also introduce another postulate, which is only apparently irreconcilable with the former, namely, that light is always propagated in empty space with a definite velocity c which is independent of the state of motion of the emitting body. These two postulates suffice for the attainment of a simple and consistent theory of the electrodynamics of moving bodies based on Maxwell's theory for stationary bodies. The introduction of a "luminiferous ether" will prove to be superfluous inasmuch as the view here to be developed will not require an "absolutely stationary space" provided with special properties, nor assign a velocity-vector to a point of the empty space in which electromagnetic processes take place.[6]

Some commentary will make the meaning of these paragraphs clearer. In a significant sense, one can talk about a relativity principle within Galilean and Newtonian mechanics. Recall that Galileo argued that there is no experiment that will allow one to detect whether one is stationary or moving at a constant velocity. Galileo, who presented this claim in reference to an enclosed cabin on a ship, thereby made a major step toward the discovery of the law of inertia. The laws of Newtonian mechanics apply for any inertial frame of reference. This is to say they apply in (and only in) those frames of reference in which bodies obey the law of inertia. For example, Newton's second law ($F = kma$) should apply equally to motions encountered either on a stationary platform, or on one moving at constant velocity.

On the other hand, if the frame of reference is accelerating, then the Newtonian laws do not in general apply. In this sense, it appears that there is no privileged frame of reference in Newtonian mechanics. Newton himself believed that the fixed stars could be used as a reference frame, which would allow him to speak in terms of absolute space, but his arguments were challenged. In fact, there seems to be no test that will allow us to determine whether the Earth is fixed in position or the Sun is, or that neither is. In this sense, a relativity principle is inherent in classical mechanics.

The situation seems different in regard to electromagnetic phenomena, for example, light. The wave theory of light, according to which light is an electromagnetic wave in the aether, implies that a fixed frame of reference exists. That fixed frame of reference is the aether. In principle, at least, we should be able to determine whether we are moving in relation to the aether,

---

[6]Albert Einstein, "On the Electrodynamics of Moving Bodies" in A. Einstein, H. A. Lorentz, H. Weyl, and H. Minkowski, *The Principle of Relativity,* a collection of classic papers trans. by W. Perrett and G. B. Jeffery (Dover republication of the 1923 original), pp. 37–38.

just as we can determine the speed of a boat relative to the lake on which it is moving. It was this conviction that motivated Michelson and Morley to carry out their experiment.

In one sense, this is the point that Einstein is discussing in his first paragraph. He refers in it to two different interpretations of a well known phenomenon concerning magnets and conducting wires. Michael Faraday had discovered that if a magnet is inserted into a loop of wire, a current is induced in the wire. Electrical theorists

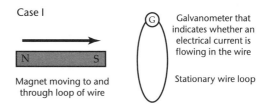

Case I

Magnet moving to and
through loop of wire

Galvanometer that
indicates whether an
electrical current is
flowing in the wire

Stationary wire loop

in the tradition of Maxwell explained this by saying that a *moving* magnet generates an electrical field, which field causes a current to flow in the wire. You might think that this result provides a test as to whether a magnet is moving or at rest: if it is at rest, no electrical field surrounds it, whereas if it is in motion, an electrical field does surround it. Thus it might seem that for electromagnetic phenomena (unlike the phenomena of mechanics), we should be able to find a fixed frame of reference, that is, a way of determining absolute motion or rest.

The following phenomenon, however, also occurs: if a conducting wire is moved along a stationary magnet, a current is induced in the wire. Maxwellians interpreted this in terms of the idea that when electrical charges in the wire move through a magnetic field, they experi-ence force, which produces a current. Experiment re-

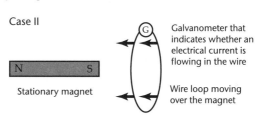

Case II

Stationary magnet

Galvanometer that
indicates whether an
electrical current is
flowing in the wire

Wire loop moving
over the magnet

veals that if the velocity of the magnet (when the wire is stationary) is equal to the velocity of the wire (when the magnet is stationary), the same current results. This precludes the possibility of determining whether the magnet or the wire is moving. It is as if nature conspires to prevent us from determining whether the magnet is fixed in position or moving. Thus it seems that a relativity principle governs not only mechanics but also electrodynamics. If this is the case, the aether cannot supply a fixed reference frame.

It was precisely such considerations as this that led Einstein to pro-pose that a single relativity principle governs both mechanics and electrody-namics—that nature, in effect, always behaves in such a way that we cannot ever determine (either in mechanics *or* in electrodynamics) a fixed frame of reference. We simply have no means whatsoever of determining whether any given body, be it apple or planet, magnet or conducting wire, is in a state

of absolute rest. Once this principle was adopted it entailed, as we shall
see, conceptual modifications not only in electrodynamics but in mechanics
as well.

The same point was made by Einstein in a different way in his *Autobiography* when he discussed a "paradox upon which I had hit at the age of
sixteen. . . ."[7] Einstein wrote:

> If I pursue a beam of light with the velocity c..., I should observe
> such a beam of light as a spatially oscillatory electromagnetic field at
> rest. However, there appears to be no such thing, whether on the
> basis of experience or according to Maxwell's equations. From the
> very beginning it appeared to me intuitively clear that, judged from
> the standpoint of such an observer, everything would have to happen
> according to the same laws as for an observer who, relative to the
> earth, was at rest.[8]

Although Einstein urged the abandonment of these fundamental ideas
of traditional physics, he proposed in the second paragraph of the passage
cited from his 1905 paper two new fundamental principles or postulates. He
designated the first of these his **"Principle of Relativity,"** which he formulated in the following way: ". . . the same laws of electrodynamics and optics
[are] valid for all frames of reference for which the equations of mechanics
hold good." Another formulation is: "The laws of nature must be the same for
any two inertial systems moving in relation to each other." Einstein's second
postulate as expressed in his second paragraph is: ". . . light is always propagated in empty space with a definite velocity c which is independent of the
state of motion of the emitting body." We shall see that these two postulates
lead to quite definite and surprising conclusions about how nature behaves.

Einstein on "Inner Perfection" and "External Confirmation"

It would be a mistake to assume that empirical considerations (if the above
can be so described) were the sole factors that helped lead Einstein to
his theory or that formed its sole support. Rather in his *Autobiography* he
stressed the importance not only of the "external confirmation" of theories
but also their "inner perfection." In particular, he wrote:

> Before I enter upon a critique of mechanics as the foundation of
> physics, something of a broadly general nature will first have to be
> said concerning the points of view according to which it is possible to
> criticize physical theories at all. The first point of view is obvious: the

---

[7] Einstein, *Autobiography,* p. 53.
[8] Einstein, *Autobiography,* p. 53.

theory must not contradict empirical facts. However evident this demand may in the first place appear, its application turns out to be quite delicate. For it is often, perhaps even always, possible to adhere to a general theoretical foundation by securing the adaptation of the theory to the facts by means of artificial additional assumptions. In any case, however, this first point of view is concerned with the confirmation of the theoretical foundation by the available empirical facts.

The second point of view is not concerned with the relation to the material of observation but with the premises of the theory itself, with what may briefly but vaguely be characterized as the "naturalness" or "logical simplicity" of the premises (of the basic concepts and of the relations between these which are taken as a basis). This point of view, an exact formulation of which meets with great difficulties, has played an important role in the selection and evaluation of theories since time immemorial. The problem here is not simply one of a kind of enumeration of the logically independent premises (if anything like this were at all unequivocally possible), but that of a kind of reciprocal weighing of incommensurable qualities. Furthermore, among theories of equally "simple" foundation that one is to be taken as superior which most sharply delimits the qualities of systems in the abstract (i.e., contains the most definite claims). Of the "realm" of theories I need not speak here, inasmuch as we are confining ourselves to such theories whose object is the totality of all physical appearances. The second point of view may briefly be characterized as concerning itself with the "inner perfection" of the theory, whereas the first point of view refers to the "external confirmation." The following I reckon as also belonging to the "inner perfection" of a theory: We prize a theory more highly if, from the logical standpoint, it is not the result of an arbitrary choice among theories which, among themselves, are of equal value and analogously constructed.[9]

Einstein's Two Postulates and a Derivation of the Special Theory of Relativity

Einstein's special theory of relativity was based on two postulates. They may be stated as follows:

1. **The Principle of Relativity:** The laws of nature must be the same for any two inertial systems moving in relation to each other.

2. The velocity of light in empty space always has the same value, independently of the motion of the light source or of the receiver of the light.

---

[9]Einstein, "Autobiography," pp. 21–23.

These postulates require the formulation of a new set of transformation equations to replace the Galilean transformation equations. Let us proceed to develop these equations, beginning from the second postulate.

If a beam of light leaves the origin O of the K frame of reference at time $t = 0$ (see Ch. 5 pp. 260–261), then we can describe its motion by means of the equation

$$x = ct \quad \rightarrow \quad x - ct = 0.$$

By our second postulate, the speed of the ray in the K' frame must also be c. Hence a beam of light originating at O' will obey the following equation:

$$x' = ct' \quad \rightarrow \quad x' - ct' = 0.$$

It is important to note that the same c occurs in both equations. Consider now the equation

$$\left(x' - ct'\right) = \lambda \left(x - ct\right),$$

where λ is a constant not equal to 0. Notice what this equation stipulates. It says that if $(x' - ct') = 0$, then it must be that $(x - ct) = 0$, and vice versa. Thus this equation summarizes the second postulate.

Now consider rays going to the left, i.e., in the negative direction. We then get the equations:

$$-x = ct \quad \rightarrow \quad x + ct = 0 \quad \text{and} \quad -x' = ct' \quad \rightarrow \quad x' + ct' = 0.$$

Combining these two equations, we get:

$$\left(x' + ct'\right) = \mu \left(x + ct\right),$$

where μ is a constant not equal to 0. Let us now combine these two equations, first by adding them, then by subtracting the second from the first. Adding them, we get:

$$
\begin{aligned}
x' - ct' &= \lambda \left(x - ct\right) \\
x' + ct' &= \mu \left(x + ct\right) \\
\hline
2x' &= (\lambda + \mu)\, x - (\lambda - \mu)\, ct.
\end{aligned}
$$

Dividing by 2, we have:

$$x' = \left(\frac{\lambda + \mu}{2}\right) \cdot x - \left(\frac{\lambda - \mu}{2}\right) \cdot ct.$$

Let $\left(\frac{\lambda+\mu}{2}\right) = a$ and let $\left(\frac{\lambda-\mu}{2}\right) = b$.
Then

$$\boxed{x' = ax - bct} \tag{1}$$

We now wish to subtract the second of the above equations from the first; the second equation becomes

$$- \left(x' + ct' \right) = -\mu \left(x + ct \right),$$

which gives:

$$
\begin{array}{rcl}
(x' - ct') & = & \lambda \left(x - ct \right) \\
- (x' + ct') & = & -\mu \left(x - ct \right) \\
\hline
-2ct' & = & \left(\lambda - \mu \right) x - \left(\lambda + \mu \right) ct.
\end{array}
$$

Dividing by 2, we have:

$$-ct' = \left(\frac{\lambda - \mu}{2} \right) x - \left(\frac{\lambda + \mu}{2} \right) ct.$$

Then

$$\boxed{-ct' = bx - act} \qquad (2)$$

We shall now attempt to determine values for the constants a and b. Take the point $x' = 0$ in the K' frame and substitute this value for x' into equation (1). We get: $0 = ax - bct$. Solving for x:

$$x = \frac{bc}{a} \cdot t.$$

Note the form of this equation. It gives the value of x, the distance of the origin of the K' frame from the origin in the K frame, as a function of t (time). This implies that bc/a must be a velocity, v, the velocity of the origin of the K' frame as measured in the K frame. Hence, we can write:

$$v = \frac{bc}{a}.$$

Imagine a meter stick in the K' frame. We wish to use our equations to calculate its length relative to the K frame. For K',

$$\Delta x' = x'_2 - x'_1 = 1 \text{ meter}.$$

The symbol Δx, which is read "delta x," simply means the difference between the two values of x. For K, from equation (1), above, with t taken as 0, i.e., with the coordinate systems coinciding, we have:

$$x' = ax - bc \cdot 0 = ax.$$

Hence

$$\Delta x' = x'_2 - x'_1 = 1 \text{ meter} = ax_2 - ax_1 = a\Delta x.$$

In short, $\Delta x' = a\Delta x$. But because $\Delta x' = 1$ meter, we have $1 = a\Delta x$, from which we get:

$$\Delta x = \frac{1}{a}.$$

Hence the meter stick in K', when viewed from K, is altered in length by a factor $1/a$.

At this point, we need to make use of Einstein's **first postulate**, which stipulates that our frames of reference must be equivalent; this means that if an observer in the K frame sees a meter stick in K' as shortened, then so also must an observer in the K' frame see a meter stick in K as shortened, and by the same amount. Let us now use our equations. Take the equation $x' = ax - bct$ and solve for t:

$$t = \frac{ax - x'}{bc}.$$

In direct analogy to what was done before, let us set $t' = 0$ in equation (2), above; this gives $act = bx$. Substituting the value derived for t into this equation, we get:

$$bx = act = \frac{ac\,(ax - x')}{bc} = \frac{a^2x}{b} - \frac{ax'}{b}$$

or

$$bx = \frac{a^2x}{b} - \frac{ax'}{b}.$$

Solving for x' gives:

$$x' = \left(\frac{a^2x}{b} - bx\right) \cdot \left(\frac{b}{a}\right) = a\left(1 - \frac{b^2}{a^2}\right)x.$$

We know that $v = \frac{bc}{a}$ and from this it is obvious that $\frac{v}{c} = \frac{b}{a}$. Hence

$$x' = a\left(1 - \frac{v^2}{c^2}\right)x.$$

This means that a meter stick in the K frame observed by a person in the K' frame will have length $a\left(1 - \frac{v^2}{c^2}\right) \cdot 1$ meter. In general, then, a meter stick in K' measured from K will have the length $\Delta x = \frac{1}{a}$ meters (as shown earlier), and, similarly, a meter stick in K observed from K' will have its length altered according to the equation $\Delta x' = a\left(1 - \frac{v^2}{c^2}\right)$ meters. Our first postulate, however, tells us that these alterations must be quantitatively identical, that is, $\frac{1}{a} = a\left(1 - \frac{v^2}{c^2}\right)$. Let us solve this for a.

$$a^2 = \frac{1}{1 - \frac{v^2}{c^2}},$$

which implies that

$$a = \frac{1}{\sqrt{1 - \frac{v^2}{c^2}}}$$

Using the value we now have for a, we can find the value of b from the equation $v = bc/a$:

$$v = \frac{bc}{a} \quad \rightarrow \quad b = \frac{av}{c} \quad \rightarrow \quad b = \frac{1}{\sqrt{1 - \frac{v^2}{c^2}}} \cdot \frac{v}{c}.$$

If we now put these values into our original equations and recall that the y- and z-coordinates are not affected by a motion perpendicular to them, we get:

$$x' = \frac{x - vt}{\sqrt{1 - \frac{v^2}{c^2}}}$$

$$y' = y$$
$$z' = z$$

$$t' = \frac{t - \left(\frac{v}{c^2}\right)x}{\sqrt{1 - \frac{v^2}{c^2}}}$$

These four equations are the **Special Theory of Relativity Transformation Equations**. As you can see, these equations correspond to the Lorentz-FitzGerald contraction. They dictate that distance in one frame of reference will be contracted when seen from the other frame by a factor of $\sqrt{1 - v^2/c^2}$, where v is the velocity of the frame that is moving with reference to the other.

Time Dilation

Let us now apply these equations to time. In the K' frame, take a clock that beats once each second, i.e., $t_2 - t_1 = 1$ second. Let the clock be located at the origin of the K' frame, i.e., at $x' = 0$. Then from the first transformation equation, we have:

$$x' = 0 = \frac{x - vt}{\sqrt{1 - \frac{v^2}{c^2}}},$$

which implies that $x = vt$.

Now take the fourth transformation equation:

$$t' = \frac{t - \left(\frac{v}{c^2}\right) x}{\sqrt{1 - \frac{v^2}{c^2}}}.$$

Let us now substitute our value for x:

$$t_2' - t_1' = 1 \text{ second} = \frac{t_2 - \left(\frac{v}{c^2}\right) v t_2}{\sqrt{1 - \frac{v^2}{c^2}}} - \frac{t_1 - \left(\frac{v}{c^2}\right) v t_1}{\sqrt{1 - \frac{v^2}{c^2}}}.$$

Hence

$$1 \text{ second} = \left(\frac{1}{\sqrt{1 - \frac{v^2}{c^2}}}\right) \cdot \left[t_2 \left(1 - \frac{v^2}{c^2}\right) - t_1 \left(1 - \frac{v^2}{c^2}\right)\right]$$

$$= (t_2 - t_1) \cdot \sqrt{1 - \frac{v^2}{c^2}}.$$

Thus

$$t_2 - t_1 = \frac{1 \text{ second}}{\sqrt{1 - \frac{v^2}{c^2}}}.$$

From the fact that $\sqrt{1 - \frac{v^2}{c^2}}$ must be less than 1, we see that

$$t_2 - t_1 > 1 \text{ second}.$$

Put in physical terms, this means that time will be expanded: the time be-tween two beats of the clock moving with K' as seen from K will be longer than the time between beats as perceived from K'. This is a **time dilation**. Note that the beatings of the clocks will not be simultaneous. Simultaneity will be **relative** to the observer's frame of reference.

Summary

Let us now review what Einstein has done, considering it within the con-text of his claim that a good theory should have both "external confirmation" and "inner perfection."[10] Basing his development on two postulates and a moderate dose of fairly elementary mathematics, Einstein has derived a number of startling physical implications. He asserts, for example, that if, while standing in a train station, we observe a meter stick and clock aboard

---

[10]Einstein, "Autobiography," pp. 21–23, quoted above.

a passing train, then we should find that the meter stick in the train is shorter than meter sticks in the station and that the time between ticks of the clock on the train is greater than the time between ticks in the station clock. Admittedly, these differences are very small; nonetheless, Einstein asserts that they do actually occur. Moreover, it is noteworthy that Einsteinian physics can account for the null result in the Michelson-Morley experiment, whereas Newtonian physics cannot. A large number of other empirical results also support Einstein's physics. In short, Einstein's system has impressive external confirmation.

Significant as such external confirmations are, Einstein would want us to recognize that quite another aspect of his theory should be seen as providing assurance of its correctness. This is what he calls its "inner perfection." An excellent example of what he has in mind by his theory's inner perfection is that he succeeded in deriving these startling physical results from nothing more than two rather simple and innocuous postulates.

The Twin Paradox

Consider identical twins, one of whom goes on a long rocket trip at a high velocity. Time in this twin's frame of reference, as seen from the Earth, will slow down so that he should, it seems, return to Earth as a younger person than his twin. Consider, however, the situation from his point of view; he will, in effect, have seen his terrestrial twin moving away from him at a large velocity. Thus it must be the case that he will perceive clocks on Earth as beating more slowly than clocks on his rocket ship and, correspondingly, he should find that his Earth-bound twin will be younger than he is! But this seems contradictory; in fact, it appears to violate Einstein's first postulate. This paradox is known as the twin paradox.

The resolution of the paradox is that these are not inertial frames. The rocket ship has been accelerating in relation to the Earth. Therefore, there is no violation of the first postulate, which applies only to reference frames moving at constant velocity relative to each other.

Four Dimensions

From the four transformation equations, the following generalized equation can be derived:

$$x'^2 + y'^2 + z'^2 - c^2 t'^2 = x^2 + y^2 + z^2 - c^2 t^2.$$

From this we see that to specify the location of an event in K in such a way as to derive its coordinates in K' requires specification not only of its x-, y-, and z-coordinates, but also of its time. This is to say that **space is four**

dimensional. To specify precisely where an object is located, one must give not only its position in the traditional meaning of that word, but also its time coordinate.

Derivation of the Equation $E = mc^2$

The most famous equation derived by Einstein is the equation $E = mc^2$, where E is energy, m is mass, and c is the speed of light. This equation states that a quantity m of matter can be completely converted into a quantity of energy E, and vice versa, with c^2 being the proportionality constant. This equation was implicit in Einstein's 1905 special theory paper and was made explicit in a short paper he published later that year.[11]

We shall derive the equation $E = mc^2$ from one of the results that Einstein derived as part of his special theory of relativity. That result is:

$$m = \frac{m_0}{\sqrt{1 - \frac{v^2}{c^2}}},$$

where m_0 is the rest mass of a body, i.e., the mass possessed by a body when at rest in a particular frame of reference. This equation implies that the mass of a body increases as the body's speed increases. Moreover, as its speed approaches that of light, the denominator in this equation approaches 0, which indicates that m, the mass of the moving body, approaches infinity. Some empirical evidence for this equation had become available even before 1905. Walter Kaufmann, a German physicist, had shown in 1902 that high speed electrons emitted from radioactive nuclei seem to have greater mass than electrons at rest.

Let us introduce the abbreviation that $\beta = \frac{v}{c}$; then our equation becomes

$$m = m_0 \left(1 - \beta^2\right)^{-\frac{1}{2}}.$$

By a well known algebraic theorem called the binomial theorem,[12] we can write that

$$\left(1 - \beta^2\right)^{-\frac{1}{2}} = 1 + \frac{\beta^2}{2} + \frac{3\beta^4}{8} + \cdots$$

---

[11] Albert Einstein, "Ist die Trägheit eines Körpers von seinem Energieinhalt abhängig?" *Annalen der Physik, 18* (1905), 639–641.

[12] The binomial theorem is generally stated in the following form:

$$(x + y)^n = x^n + nx^{n-1}y + \left[\frac{n(n-1)}{2!}\right]x^{n-2}y^2 + \left[\frac{n(n-1)(n-2)}{3!}\right]x^{n-3}y^3 + \cdots + y^n.$$

For small values of β, however, terms involving β to the fourth or higher powers become negligibly small. Thus we can write:

$$m = m_0 \left(1 + \frac{\beta^2}{2}\right).$$

Thus we have:

$$m = m_0 \left(1 + \frac{1}{2}\frac{v^2}{c^2}\right) = m_0 + \frac{1}{2}m_0\frac{v^2}{c^2}.$$

Let us use this formula to compute the change in mass Δm when a body of rest mass m_0 goes from rest to velocity v. We see that

$$\Delta m = m - m_0 = \left(m_0 + \frac{1}{2}m_0\frac{v^2}{c^2}\right) - m_0$$
$$= \frac{1}{2}m_0\frac{v^2}{c^2}.$$

As noted previously, $\frac{1}{2}m_0v^2$, classically understood, is the kinetic energy of a body with constant mass m_0 and moving with velocity v. The relativistic change in mass Δm is thus equal to the kinetic energy that was added to the body to put it in motion, divided by c^2. This suggests that the relativistic change in mass is actually a change in kinetic energy—that the kinetic energy added to the body manifests itself as increased mass. Although this example involves kinetic energy specifically, one can show generally that energy E in any form can be seen as equivalent to mass m, where $m = E/c^2$, or $E = mc^2$.

For example, compressing a spring adds energy (potential energy) to it and thereby increases its mass. The same is true if heat energy is added to a rod; the rod increases in mass. Typically, the change in mass will be extremely small. The equation $E = mc^2$ required a reformulation of two of the most fundamental laws of physics, the laws of conservation of matter and conservation of energy. They were replaced by the single law of the **joint conservation of energy and matter**.

Information on quantitative aspects of this famous equation provides an insight into its significance. For example, it has been calculated that to produce the quantity of energy equal to that generated each day by all our traditional sources of energy such as hydroelectric generators, internal combustion engines, and steam engines, would require the complete destruction of only about 10 grams of nuclear fuel.

The General Theory of Relativity

In 1916, Einstein published what is known as his general theory of relativity. The following presentation of it is based primarily on the formulation given by Einstein and Infeld in their *Evolution of Physics,* pp. 209–45. Because the general theory of relativity is quite complex, we shall be far less able to get at its fundamental ideas than was the case for the special theory of relativity.

One of the fundamental claims in the general theory is that the inertial mass and the gravitational mass of a body are **necessarily** equivalent. This equivalence is numerically true in Newtonian mechanics, but is viewed as a fortunate accident. The **inertial mass** of a body is that feature of it that makes it resist changes in motion. The inertial mass of a car makes it difficult to get it moving and, once it is moving, to get it stopped. The **gravitational mass** is that aspect of the body that determines the gravitational force on it. Recall that in Chapter 5 (pp. 247–248) it was shown that if inertial mass M_i and gravitational mass M_g are numerically equal, then if we combine Newton's second law of motion with his law of gravitation,

$$F = M_i\, a = \frac{GM_g\, M'_g}{D^2},$$

the two masses M_i and M_g mutually cancel. We have

$$a = \frac{GM'_g}{D^2},$$

showing that all bodies at the same distance D from a gravitating body M'_g will experience the same gravitational acceleration a, independent of their masses.

That inertial and gravitational mass are to be seen as necessarily equivalent becomes clear from the following thought experiment. Suppose you are in a windowless elevator. Let us also assume that you discover that in that elevator, objects remain exactly where they are placed: a picture put on the wall stays there, even without a hook; an object when thrown continues to move in a straight line. What could be causing objects to behave in such a way? Two possible explanations come to mind:

1. No gravitational forces act on objects in the elevator.

2. The elevator is in free fall; for example, its cable has broken.

Suppose now that you are located in an elevator in which objects behave just as they do on Earth. Coins, when dropped, fall to the ground; when thrown, they follow parabolic paths. How is this to be explained? Again two explanations come to mind.

1. The elevator is located in a gravitational field that influences the motions of the objects in the elevator.

2. The elevator has no gravitational forces acting on it, but is being hauled up with a constant acceleration by a cable attached to a hook in the roof of the elevator. In this case, the coin remains at rest (since no gravitational force is acting on it), but this means that it appears to fall and will collide with the floor of the elevator.

Given these two possibilities, how can one decide which is the correct explanation? Let us turn to what is called Einstein's **Principle of Equivalence**, one formulation of which is: In a small region of space-time, it is not possible operationally to distinguish between a frame at rest in a uniform gravitational field and a frame being uniformly accelerated in empty space. This principle stipulates that there cannot be any observational or experimental means that permit us to distinguish between these two explanations. But if we cannot distinguish between a body being gravitationally attracted and a body undergoing mechanical acceleration, then we cannot distinguish between gravitational and inertial mass.

The principle thus entails that gravitational mass and inertial mass are equivalent, that $M_g = M_i$. It also entails some surprising new effects. For example, suppose a ray of light enters a window in one side of the elevator. If the elevator is motionless in a gravitational field, we would ordinarily expect the ray to move in a straight line whereas if the elevator is being drawn up, we would expect the ray to appear to curve toward the floor.

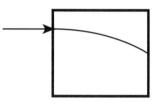

If, however, Einstein's principle of equivalence is correct, the ray must follow a curved path in both cases—in the gravitational field as well as in the accelerating elevator. We shall see shortly that such a result is confirmed by actual observation. But first let us see what the general theory of relativity implies for space-time.

In the general theory of relativity, both the traditional view of space as Euclidean and the traditional notion of gravitation must be given up. For Einstein, the presence of matter at any point distorts space-time in its vicinity. Near massive bodies, projectiles do not follow straight lines, but rather curved lines determined by distortions produced by the massive body.

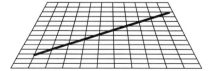
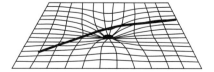

The Three Classic Tests of the General Theory of Relativity

In presenting Einstein's general theory of relativity, mention is traditionally made of three classic tests. These are described below, but it is important to note that at least two other types of evidence also support it. First, because all results obtained in classical (i.e., pre-1900) physics can be explained by the general theory of relativity, the success of classical physics provides evidence for the new theory. Second, Einstein repeatedly stressed that the inner perfection of a theory constitutes evidence for it. (See, for example, the previously-quoted section of his *Autobiography*). And he believed that the general theory of relativity possesses greater inner perfection than Newtonian physics. For example, in Einstein's general theory the numerical equivalence of inertial and gravitational mass is a consequence of the fundamental principles, rather than being a happy accident needing empirical verification, as it was in Newtonian physics.

First Test: The Excess in the Advance of the Perihelion of Mercury

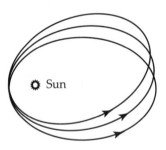

According to Kepler's first law of motion, the planet Mercury moves in an ellipse. It turns out, however, that this ellipse precesses; that is, its major axis gradually shifts (see diagram), although it always passes through the Sun. This means that the perihelion of Mercury's orbit (the point where Mercury passes nearest the Sun) gradually moves around a circle; the rate of its motion is 574 seconds of arc per century. This is a small but observable amount.

Classical mechanics could account for most of this amount; in particular, 531 of the 574 seconds could be explained as due to the effects of the other planets on Mercury's orbit. This left, however, **43 seconds of arc** unexplained. In the 1850s, the French astronomer Leverrier postulated the existence of one or more planets orbiting inside the orbit of Mercury as a means of explaining this excess. In this way, he was able to account for most of the effect. Despite numerous searches and a couple dozen supposed sightings of the postulated planet (sometimes called "Vulcan"), the object was never in fact found.

Although Einstein was unaware of this problem when he began to formulate his general theory, he found in 1915 that his new theory would account for this effect. At that time, he wrote Arnold Sommerfeld:

> This last month I have lived through the most exciting and the most exacting period of my life; and it would be true to say that it has been

the most fruitful. . . . The wonderful thing that happened. . . was that not only did Newton's theory result from it [the general theory of relativity], *as a first approximation,* but also the perihelion motion of Mercury, *as a second approximation.*[13]

Einstein proceeded to publish this result in a 1915 paper and was able to cite it when in 1916 he published his overall presentation of his general theory of relativity. He believed that a similar effect is present in the other planets, but that it had not been detected because the other planets, being farther from the Sun, are less influenced by the strong distortion of space-time in the vicinity of the Sun. It has subsequently been detected for Venus, the Earth, and the minor planet Icarus.

Second Test: The Eclipse Test

As we have seen, the general theory of relativity predicts a bending of light rays in the vicinity of massive bodies. At first Einstein feared that the effect would be observationally undetectable. In a 1911 paper, Einstein predicted that light rays passing by the Sun should exhibit a bending of .83 seconds of arc. The astronomer Erwin Freundlich planned to check this prediction during a solar eclipse due to be visible in central Russia on 21 August 1914, but he was prevented from doing so by the onset of World War I. In a 1915 paper on the advance of the perihelion of Mercury, Einstein announced a new determination of this amount, predicting a bending of 1.75 seconds of arc, double what he had predicted in 1911. This prediction was tested during a solar eclipse on 29 March 1919. Because the Moon then blocked the light of the Sun, it became possible to compare the observed positions of stars nearly collinear with the Sun to photographs of the same stellar region with the Sun absent. Britain sent out eclipse expeditions to Africa and to Brazil. The resulting determination was announced at a special joint meeting of the Royal Society and the Royal Astronomical Society held on 6 November 1919. A shift of 1.64 seconds was reported, thus strongly supporting Einstein, whose international fame can be dated from that striking confirmation of his theory. At subsequent eclipses, additional measurements have been made. Within observational error, all agreed with Einstein's prediction.

Third Test: The Gravitational Red Shift

Another conclusion of the general theory of relativity is that time should be slowed in the vicinity of massive bodies; for example, a clock near a massive

---

[13]As quoted in C. P. Snow, "Albert Einstein, 1879–1955" in A. P. French (ed.), *Einstein: A Centenary Volume* (Harvard University Press, 1979), p. 4.

body should beat more slowly than an identical clock elsewhere. The rate at which light is emitted from atoms is dependent on time, which entails that radiating atoms in effect function as clocks. In particular, the frequency of the light rays emitted by atoms in strong gravitational fields should be reduced. Thus rays should be shifted toward the low frequency (red) end of the spectrum. As in the other tests, this **red shift** was predicted to be very small. It was hoped that it might be detected through observations of such stars as Sirius B, a companion of the very bright star Sirius. Sirius B has a density about 30,000 times greater than water. Although this suggests that in principle this effect could be determined, good results from these and comparable tests began to be obtained only during the 1960s.[14] The concept of a relativistic red shift has been used in discussing the gravitational collapse of very dense stars. Because in extremely dense stars, the frequency tends to shift toward zero, no light is seen. Such stars are called "black holes."

Conclusion

The general theory of relativity is not as strongly supported as the special theory; in fact, other theories entailing predictions that differ from those of Einstein by only very small amounts have been developed. Nonetheless, it remains a highly respected and fundamental theory in present day theoretical physics.

One of the most important conclusions that should have become clear from this discussion of Einstein's special and general theories of relativity is how seriously mistaken is the widely repeated claim that Einstein proved that "everything is relative." Einstein, no less than his great predecessors Aristotle, Galileo, and Newton, incorporated a number of absolutes into his system. One example of this is that in the special theory of relativity, Einstein specified that the velocity of light is an absolute quantity. And another of his absolutes is, ironically, the principle of relativity itself.

A Final Comment Concerning the Philosophies of Science of Mach, Planck, and Einstein

Albert Einstein and Max Planck (1858–1947) are widely regarded as the two leading physicists of the early twentieth century. At approximately the

---

[14]The gravitational red shift should not be confused with the Doppler shift, which results from the relative velocity of bodies moving towards or away from the Earth.

same time that Einstein was formulating relativity theory, Planck was creating the other most fundamental physical theory to be developed in the twentieth century, the quantum theory, which is dated from a paper Planck published in 1901. The eminence of Einstein and Planck makes it a particularly interesting fact that both, early in their careers, were enthusiastic about the philosophical doctrines advocated by Ernst Mach. Mach, who although trained in physics became best known as a philosopher of science, espoused and developed **positivistic** and **empiricist** positions in philosophy. The goal of scientific theory, Mach urged, is not truth, but rather **economy of thought**. Theories are above all methods of binding together and coordinating sensations, which, Mach believed, were the only real entities in the universe. Mach's stress on empirical information led him to disparage metaphysics; indeed, he crusaded to remove metaphysical ideas from science. Among such ideas were the notions of absolute space and time and also the idea that matter is composed of atoms. Such ideas, he stressed, cannot be empirically derived; hence they have no legitimate role in science. Mach's model or paradigm science was thermodynamics. He saw thermodynamics as ideally representing what a science should be because the fundamental entities of thermodynamics (pressure, temperature, volume) are directly observable. Moreover, many of its laws were exactly what Mach thought scientific laws ought to be, namely, correlations of observable properties.[15]

By 1908, Planck had broken from Mach's views. In a famous address Planck delivered in 1908 and entitled "The Unity of the Physical Universe" (republished in Planck, *A Survey of Physical Theory*), Planck attacked the Machian philosophy in strong phrases. At one point he summarized Mach's philosophy by stating:

> Is the physical world simply a more or less arbitrary creation of the intellect, or are we forced to the opposite conclusion that it reflects phenomena which are real and quite independent of us? Expressed in a concrete form, can we rationally assert that the principle of conservation of energy was true, even when nobody could think about it, or that the heavenly bodies will move according to the law of gravitation when the earth and all that is therein is in ruin?
>
> If, looking back over the past, I give an affirmative answer to these questions, I am certain that this answer is, in a way, contradictory to a tendency of natural philosophy (recently introduced by Ernst Mach) which is in great favour in scientific circles. According to this, nothing is real except the perceptions, and all natural science is ultimately an economic adaptation of our ideas to our perceptions, to which we are

---

[15] An example is Boyle's Law: that the volume of a gas is inversely proportional to the pressure on it.

driven by the fight for existence. The boundary between physical and psychical research is only practical and conventional. The only real elements of the world are the perceptions.[16]

Turning then to a specific area of contention, Planck expressed some very un-Machian views:

> ... I would expressly emphasize that the arguments which have been directed from every side against the atomic hypothesis and the electron theory are unjustifiable and unwarrantable. I would even like to assert—and I know I am not alone in this—that atoms, little as we know of their actual properties, are as real as the heavenly bodies, or as earthly objects around us....[17]

Later in his address Planck cited a historical argument on behalf of the realist position, which he was advocating in opposition to Mach and the positivists. Planck stated his argument in the following way:

> As the great masters of exact research threw out ideas in science—as Copernicus removed the centre of the universe from the earth, as Keppler [sic] propounded the laws named after him, as Newton discovered general gravitation, as Huygens set forth the undulatory theory of light, as Faraday created the foundations of electro-dynamics—very many more can be cited—the economic [i.e., Machian] point of view was the very last with which these men armed themselves in the war against inherited opinions and commanding authority. No—they were moved by their fixed belief in the reality of their picture, whether founded on an intellectual or a religious basis.[18]

That Planck felt strongly about this issue is clear from his final paragraph where he recommended "a lasting confidence in the force of the Word which, for more than 1900 years, has given us an ultimate infallible test for distinguishing false prophets from true—'By their fruits ye shall know them!'"[19] And throughout the remainder of his life, Planck continued to advocate a realist position.

The paths of Planck and Einstein crossed on many occasions. For example, it was Planck who accepted for publication Einstein's 1905 paper on the special theory of relativity; moreover, shortly thereafter, Planck taught a seminar on it. On the other hand, Einstein's 1905 paper on the

---

[16] Max Planck, "The Unity of the Physical Universe" in Planck's *A Survey of Physical Theory*, trans. R. Jones and D. H. Williams (Dover, 1960), p. 22.
[17] Planck, "Unity," p. 23.
[18] Planck, "Unity," p. 25.
[19] Planck, "Unity," p. 26.

photoelectric effect is chiefly notable because he showed how Planck's quantum theory could be applied to explain the photoelectric effect. Although Planck had broken from Mach by 1908, Einstein remained enthusiastic about Mach's teachings.   As noted previously, Einstein's reading of Mach's historical study of mechanics, which contained a critique of Newton's views on absolute space and time, had helped Einstein to the special theory of relativity.   Einstein's continuing enthusiasm for Mach is shown by a 1909 letter that Einstein wrote to Mach, signing the letter: "Your respectful student, Einstein."[20] In 1913, as Einstein was formulating his general theory of relativity and his prediction of the bending of light rays, he wrote to Mach, stating: "your inspired investigations into the foundation of mechanics—despite Planck's unjust criticism—will [thereby] receive a splendid confirmation. . . ."[21] Mach in 1909 had publicly expressed his acceptance of relativity theory, but in 1913, he backed off from that endorsement in the preface to a book that was published only in 1922, four years after Mach had died. Earlier that year, Einstein's disaffection from Machian positivism became partly evident when in a conference in Paris, Einstein noted that Mach's philosophy could provide only a catalogue, not a system.

By 1929 it was becoming known, as one physicist put it, that "Einstein was entirely in accord with Planck's view that physical laws describe a reality in space and time that is independent of ourselves."[22] In a letter to Moritz Schlick, the founder of a famous positivist group known as the Vienna Circle, Einstein described himself as a metaphysician and added that all men are also metaphysicians. In 1931, when Einstein visited the California Institute of Technology, he made it clear—to the distress of the empiricists—that the Michelson-Morley experiment had played little part in his development of the special theory of relativity. At about the same time, Einstein began an essay he wrote on Maxwell with the statement: "The belief in an external world independent of the perceiving subject is the basis of all natural science."[23]

Einstein's break from empiricism was evident at other points in his publications from the 1930s. In 1933, Einstein stated: "the axiomatic bases of theoretical physics cannot be extracted from experience but must be freely invented. . . ."[24] Three years later he described the fundamental error of many nineteenth-century physicists as being their failure to realize that "there is no inductive method which could lead to the fundamental concepts

---

[20]As quoted in Stanley L. Jaki, *The Road of Science and the Ways to God* (University of Chicago Press, 1978), p. 182.

[21]As quoted in Jaki, *Road,* p. 182.

[22]As quoted in Jaki, *Road,* p. 185.

[23]As quoted in Jaki, *Road,* p. 186.

[24]As quoted in Jaki, *Road,* p. 404.

of physics." [25] And his developing anti-empiricist position appears in various passages in his *Evolution of Physics,* which he published in 1938.[26]

The final passage that I wish to cite is from a letter that Einstein wrote on March 30, 1952 to his friend Maurice Solovine. The passage draws together a number of fundamental features of Einstein's philosophy:

> You find it surprising that I think of the comprehensibility of the world (insofar as we are entitled to speak of such world) as a miracle or an eternal mystery. But surely, a priori, one should expect the world to be chaotic, not to be grasped by thought in any way. One might (indeed one *should*) expect that the world evidenced itself as lawful only so far as we grasp it in an orderly fashion. This would be a sort of order like the alphabetical order of words. On the other hand, the kind of order created, for example, by Newton's gravitational theory is of very different character. Even if the axioms of the theory are posited by man, the success of such a procedure supposes in the objective world a high degree of order, which we are in no way entitled to expect a priori. Therein lies the "miracle" which becomes more and more evident as our knowledge develops. And here is the weak point of positivists and of professional atheists, who feel happy because they think that they have preempted not only the world of the divine but also of the miraculous. Curiously, we have to be resigned to recognizing the "miracle" without having any legitimate way of getting any further. I have to add the last point explicitly, lest you think that weakened by age I have fallen into the hands of priests.[27]

Thus we see that both Planck and Einstein ended up advocating a realist position in science. This should not come as a surprise because, as we have seen before, Copernicus, Kepler, and Galileo were also advocates of realism.

The story we have now recounted spans almost the full spectrum of human inquiry, engaging with equal seriousness both the most fundamental and the most revolutionary ideas, demanding of us both the most freely imaginative thinking and the most carefully disciplined.

In the Preface to this book I suggested that this the most remarkable story in all secular history, supporting this claim by quoting Alfred North Whitehead's description of the developments from Galileo through Newton as humanity's "greatest single intellectual success" and citing Sir Herbert

---

[25] As quoted in Jaki, *Road,* pp. 404–405.
[26] See, for example, Einstein and Infeld, *Evolution of Physics,* p. 238.
[27] As quoted in Jaki, *Road,* pp. 192–193.

Butterfield's remark that because the scientific revolution transformed "the whole diagram of the physical universe and the very texture of human life itself, it looms so large as the real origin both of the modern world and of the modern mentality that our customary periodisation of European history has become an anachronism and an encumbrance."

Remarkable as this story and its central characters were, scarcely less remarkable is the fact that we today—presumably without similar gifts of genius and certainly with but limited time for study—have been able to enter fully into that investigation, to engage in some cases with the very words of its chief architects, to participate in the unfolding tale, and, by bringing it within our understanding, to make it our own.

The sciences of nature offer us a dual legacy: not only do they deepen our insight and sharpen our understanding of the workings of the material world, they also nourish our sense of worth and of wonder. In the story of mechanics this precious duality is especially evident. Simultaneously, we see our worth in being able to comprehend these complex issues and also enhance our wonder at the power of human thought to comprehend our cosmos.

Appendix: Galileo Laboratory

Introduction

Galileo Galilei, who frequently and famously stressed the importance of experimental investigations, had exceptional gifts as an experimentalist. This
appendix offers the reader the opportunity to carry out a set of experiments
in pendular motion using easily accessible materials. The first three experiments are quite straightforward and are no doubt comparable to those in
many manuals for physics labs. The fourth experiment, however, is more
challenging in that (1) it involves an attempt to replicate an experiment
that Galileo actually reported, and (2) if carefully performed, the experiment
should yield results that may provide some interesting historical information
as well as a surprise and a significant insight.

The materials needed for the experiments are:

- Some string, fishline, or heavy thread.

- A stopwatch and ruler.

- Three small weights, for example, weights used to hold down a fishing
  line. Alternatively, three identical coins, for example, three quarters
  (25-cent pieces) will suffice.

- Adhesive or duct tape for attaching the weights to the string.

- A stable support from which to suspend the pendulums formed from
  the weights and string. The support could be a one foot long two-by-
  four or a metal rod clamped to a sturdy table, or any comparable object
  that can be made sufficiently stable that it will not be affected by the
  swing of a pendulum suspended from it. An ideal support would be
  metal rod held in a vise.

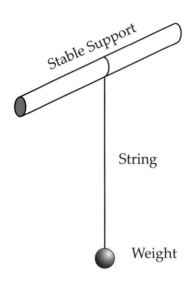

With these materials one can easily construct a device such as that shown in the accompanying diagram.

As we saw in the chapter on Galileo, he made the striking claims that the period of a pendulum's motion is independent of the weight of the pendulum and is also independent of the arc through which it is displaced. Moreover, he not only claimed that its period of motion depends on the length of the pendulum, but also specified that for pendulums of lengths L_1 and L_2, their periods will be related to their lengths according to the equation

$$\frac{T_1}{T_2} = \frac{\sqrt{L_1}}{\sqrt{L_2}}.$$

This means that a pendulum four times longer than another will swing with a period twice as long as that of the shorter pendulum. We will test each of these claims.

Experiment 1

Before beginning this experiment, reflect on what factors could influence the period of swing of a pendulum. Also reflect on what difficulties Galileo might have encountered in performing experiments that would test his claims. For example, because Galileo lived centuries before the ready availability of electronic stop watches, he had to use far more elementary devices for time measurement. A striking example of this is that for some experiments he used his pulse as a measure of equal units of time. At other points, he used water clocks.

Construct a pendulum one meter long by taping a string to a single quarter (25-cent coin) or some other small weight. The weight at the end of the string is called the bob. Measure the length of the string from the point of attachment at the top to the middle of the bob. Tape a sheet of paper to the floor below the low point of the pendulum and draw a line directly below the weight and perpendicular to the arc through which the weight swings. Sighting the vertical line from the center of the bob to the floor, displace the pendulum 10 centimeters to the left of its vertical rest position, and allow it to swing for 10 complete cycles. Use of a large carpenter's square will help in setting this up. By sighting on the line on the paper, time 10 of these cycles. (Note that one complete cycle will pass the mark *twice*.) Repeat this observation for displacements of 20, 30, and 50 cm. Be sure to divide your times by 10. Record your results below.

For 10 cm displacement, period of swing = _____

For 20 cm displacement, period of swing = _____

For 30 cm displacement, period of swing = _____

For 50 cm displacement, period of swing = _____

What conclusions can you draw from these data?

Experiment 2

Construct a heavier pendulum bob by doubling the weight (e.g., by taping together two quarters instead of one), again making the pendulum length equal to one meter. Repeat the same set of measurements for this pendulum and record your results.

For 10 cm displacement, period of swing = _____

For 20 cm displacement, period of swing = _____

For 30 cm displacement, period of swing = _____

For 50 cm displacement, period of swing = _____

What conclusions can you draw from these data? After comparing your data for this experiment with the data from Experiment 1, can you draw any additional conclusions?

Experiment 3

In this experiment you are to construct two pendulums such that when the first swings four times, the second will swing twice. Using two identical weights, make the first pendulum 20 centimeters in length and the second 80 centimeters in length, measuring from the point of suspension to the center of the bob. The two bobs should be of equal weight. Measure the periods of swing, T_1 and T_2, for these two pendulums, using the same techniques that you employed previously. At this point, we are testing Galileo's claim that the ratio of the periods (times) of swing of any two pendulums is equal to the ratio of the square roots of the lengths of their suspensions, or

$$\frac{T_1}{T_2} = \frac{\sqrt{L_1}}{\sqrt{L_2}}$$

Compare this with your own data:

Length of first pendulum $= L_1 = 20$ centimeters.

Length of second pendulum $= L_2 = 80$ centimeters.

Measured period of first pendulum $= T_1 = $ _____

Measured period of second pendulum $- T_2 - $ _____

Calculate the ratio of the square roots of the measured lengths:

$$\frac{\sqrt{L_1}}{\sqrt{L_2}} = $$ _____

Calculate the ratio of the measured periods:

$$\frac{T_1}{T_2} = $$ _____

If Galileo's equation is correct, these two ratios should be equal. If they are not equal, what do you conclude?

Experiment 4

In his *Two New Sciences,* Galileo has his spokesperson, Salviati, describe an interesting experiment with pendulums.

> Seeing that you like these novelties so well, I must show you how the eye, too, and not just the hearing, can be amused by seeing the same play that the ear hears.
>
> Hang lead balls, or similar heavy bodies, from three threads of different lengths, so that in the time that the longest makes two oscillations, the shortest makes four and the other makes three. This will happen when the longest contains sixteen spans, or other units, of which the middle [length] contains nine, and the smallest, four. Removing all these from the vertical and releasing them, an interesting interlacing of the threads will be seen, with varied meetings such that at every fourth oscillation of the longest, all three arrive unitedly at the

same terminus; and from this they depart, repeating again the same period.[1]

The goal of Experiment 4 is to test the claims Galileo made in this paragraph, in particular, to duplicate this experiment. To do this construct three pendulums, one of length 20 cm, the second of length 45 cm, and the third of length 80 cm. (Note that these lengths preserve the ratios among the lengths of the pendulums in the experiment Galileo devised, i.e., $4 \times 5 = 20$, $9 \times 5 = 45$, and $16 \times 5 = 80$). The weights can be the same in all three pendulums.

Part 1. From the information provided by Galileo, it is possible to calculate what the relationships are between the periods of these pendulums. The information given entails that for some time period X, the first will swing 4 times, the second 3 times, and the third 2 times. From this it is evident that $X = 4T_1 = 3T_2 = 2T_3$. Use the information in this equation to derive Galileo's reported results for the ratios between the periods and enter it in the first column in Table I, below. For example, from knowing that $4T_1 = 3T_2$, one can conclude that $\frac{T_1}{T_2} = \frac{3}{4}$. Enter this value at the top of the first column. Then use the same method to calculate the values for $\frac{T_1}{T_3}$ and $\frac{T_2}{T_3}$. Enter these values.

Part 2. Make your own measurements of the three periods using your pendulums of lengths 20 cm., 45 cm., and 80 cm. After getting empirical values for T_1, T_2, and T_3, calculate the ratios between these values and enter them in the second column of Table I, below.

T_1 = Period of pendulum 1 (length = 20 cm) = _____

T_2 = Period of pendulum 2 (length = 45 cm) = _____

T_3 = Period of pendulum 3 (length = 80 cm) = _____

Part 3. It is also possible to use the formula

$$\frac{T_x}{T_y} = \frac{\sqrt{L_x}}{\sqrt{L_y}}$$

to calculate the ratios between the periods. Calculate these theoretical values from the lengths of the three pendulums. Enter your results in the third column of the table.

---

[1] Galileo Galilei, *Two New Sciences, Including Centers of Gravity and Force of Percussion,* trans. Stillman Drake (University of Wisconsin Press, 1974), p. 107.

TABLE I

	Values calculated from Galileo's report	Values calculated from your measurements	Values derived theoretically from the lengths
$\dfrac{T_1}{T_2}$	_____	_____	_____
$\dfrac{T_1}{T_3}$	_____	_____	_____
$\dfrac{T_2}{T_3}$	_____	_____	_____

What level of agreement or disagreement did you obtain among the three sets of values for the ratios between the periods? If there is substantial disagreement, how would you explain such discrepancies? Should you wish for commentary on Experiment 4, you will find it at

http://www.greenlion.com/galileolab.pdf.

Selected Bibliography[1]

In General

Applebaum, Wilbur (ed.), *Encyclopedia of the Scientific Revolution* (Garland, 2000). Contains authoritative articles on many scientists, ideas, and institutions from the scientific revolution period.

Ariew, Roger, and Peter Barker, "Duhem and Continuity in the History of Science," *Revue internationale de philosophie 46* (1992), 323–343. Discusses whether the development of physical science, particularly mechanics, was a continuous or discontinuous process in the period from the middle ages to the seventeenth century.

Barbour, Julian B., *Absolute or Relative Motion: A Study from a Machian Point of View of the Discovery and the Structure of Dynamical Theories,* vol. I: *The Discovery of Dynamics* (Cambridge University Press, 1989). A very valuable study covering many aspects of mechanics.

Burtt, E. A., *The Metaphysical Foundations of Modern Physical Science* (Harcourt Brace, 1925). A classic study of the philosophical ideas of the main figures in the physical sciences during the scientific revolution.

Clagett, Marshall, *The Science of Mechanics in the Middle Ages* (University of Wisconsin Press, 1959).

Cohen, I. Bernard, *The Birth of a New Physics,* rev. and updated ed. (W. W. Norton, 1985). An excellent elementary presentation of the line of development that culminated in Newtonian mechanics. Cohen was one of the world's leading Newton scholars.

Crowe, Michael J., *Theories of the World from Antiquity to the Copernican Revolution* (Dover, 1990).

Damerow, P., G. Freudenthal, P. McLaughlin, and J. Renn, *Exploring the Limits of Preclassical Mechanics: A Study of Conceptual Development in*

---

[1] In some cases, bibliographies are designed to be an accurate record of the sources consulted by the author in writing the book. In other cases, they are designed to help readers find further materials relevant to the persons and ideas treated in the book. This bibliography seeks to combine features of both approaches.

Early Modern Science: Free Fall and Compounded Motion in the Work of Descartes, Galileo, and Beeckman (Springer-Verlag, 1991). Contains a very full bibliography.

Dijksterhuis, E. J., *The Mechanization of the World Picture* (Oxford University Press, 1961). A thorough and philosophically sophisticated history from antiquity to Newton.

Drake, Stillman, *History of Free Fall: Aristotle to Galileo* (Wall and Emerson, 1989).

Duhem, Pierre, *The Aim and Structure of Physical Theory,* trans. by Philip P. Wiener (Atheneum, 1974).

_____, _____, *Evolution of Mechanics,* trans. by Michael Cole (Sijthoff & Noordhoff, 1980).

_____, _____, *Système du monde,* 10 vols. (Hermann, 1915–1958).

_____, _____, *Medieval Cosmology: Theories of Infinity, Place, Time, Void, and the Plurality of Worlds,* trans. Roger Ariew (University of Chicago Press, 1985). Consists of a selection from Duhem's *Système du monde.*

Linton, C. M., *From Eudoxus to Einstein: A History of Mathematical Astronomy* (Cambridge University Press, 2004).

Mach, Ernst, *The Science of Mechanics,* trans. by Thomas J. McCormack (Open Court, 1942). A classic study written from a positivistic point of view.

Westfall, Richard S., *Force in Newton's Physics: The Science of Dynamics in the Seventeenth Century* (Elsevier, 1971). Written by the author of the leading biography of Newton.

Galileo

Blackwell, Richard, "Galileo Galilei" in *The History of Science and Religion in the Western Tradition: An Encyclopedia,* ed. by Gary Ferngren (Garland, 2000), 85–89.

Coelho, Victor (ed.), *Music and Science in the Age of Galileo* (Kluwer Academic Publishers, c. 1992).

Coffa, José Alberto, "Galileo's Concept of Inertia," *Physis 10* (1968), 261–281.

Drake, Stillman, "A Further Reappraisal of Impetus Theory: Buridan, Benedetti, and Galileo," *Studies in History and Philosophy of Science 7* (1976), 319–336.

_____, _____, "Galileo and the Law of Inertia," *American Journal of Physics 32* (1964), 601–608.

_____, _____, "Galileo Gleanings XVII. The Question of Circular Inertia," *Physis 10* (1968), 282–298.

_____, _____, *Galileo at Work: His Scientific Biography* (University of Chicago Press, 1978). Drake devoted his scholarly career to Galileo. Regarded as the best of a number of biographies of that great Italian scientist.

Erlichson, Herman, "Galileo's Pendulums and Planes," *Annals of Science 51* (1994), 263–272.

Fantoli, Annibale, *Galileo: For Copernicanism and for the Church* (University of Notre Dame Press, ca. 1994).

Finocchiaro, Maurice, *The Galileo Affair: A Documentary History* (University of California Press, 1989).

_____, _____, "Methodological Aspects of Galileo's Thought," *Theoria et Historia Scientiarum 2* (1992), 80–91.

Koestler, Arthur, *The Sleepwalkers; A History of Man's Changing Vision of the Universe* (Macmillan, 1959). A controversial but engaging study by a widely known author.

Koyré, Alexandre, *Galileo Studies,* trans. by John Mepham (Harvester, 1978).

Langford, Jerome J., *Galileo, Science, and the Church,* 3rd ed. (University of Michigan Press, ca. 1992). A well regarded study.

Machamer, Peter (ed.), *The Cambridge Companion to Galileo* (Cambridge University Press, 1998). Contains a number of excellent articles by experts on Galileo.

MacLachlan, James, "Galileo's Experiments with Pendulums: Real or Imaginary," *Annals of Science 33* (1967), 173–185.

McMullin, Ernan (ed.), *Galileo: Man of Science* (Basic Books, 1968; reprint: Scholar's Bookshelf, 1988). A collection of papers presented at the University of Notre Dame in 1964 to mark the 400th anniversary of Galileo's birth.

_____, _____ (ed.), *The Church and Galileo* (University of Notre Dame Press, 2005). Contains essays commenting on the relations between Galileo and the Roman Catholic Church.

Matthews, Michael, *Time for Science Education: How Teaching the History and Philosophy of Pendulum Motion Can Contribute to Science Literacy* (Plenum-Kluwer, 2000).

Naylor, R. H., "Galileo's Simple Pendulum," *Physis 16* (1974), 23–46.

Reston, James, Jr., *Galileo: A Life* (Harper Collins, 1994).

Righini-Bonelli, M. L., and W. R. Shea (eds.), *Galileo's New Science of Motion: Reason, Experiment, and Mysticism in the Scientific Revolution* (Science History Publications, 1975).

Segré, Michael, *In the Wake of Galileo* (Rutgers University Press, ca. 1991). Deals with the period after Galileo.

_____, _____, "The Role of Experiment in Galileo's Physics," *Archive for History of Exact Sciences 23* (1980), 227–252.

Settle, Thomas B., "An Experiment in the History of Science," *Science 133* (1961), 19–23.

_____, _____, "Galileo and Early Experimentation" in Rutherford Aris and H. Ted Davis (eds.), *Springs of Scientific Creativity. Essays on Founders of Modern Science* (University of Minnesota Press, 1983), 3–20.

Sharratt, Michael, *Galileo: Decisive Innovator* (Blackwell, 1994). An excellent biography. Includes information on the recent efforts at the Vatican to reassess the Galileo trial.

Westfall, Richard S., *Essays on the Trial of Galileo* (University of Notre Dame Press, 1989).

Wisan, Winifred Lovell, "Galileo's Scientific Method: A Reexamination" in Robert E. Butts and Joseph C. Pitt (eds.), *New Perspectives on Galileo* (Papers Deriving from and Related to a Workshop on Galileo held at Virginia Polytechnic Institute and State University, 1975) *University of Western Ontario Series in Philosophy of Science 14* (Reidel, 1978), 1–58.

Descartes

Ariew, Roger, "Descartes as Critic of Galileo's Scientific Methodology," *Synthese 67* (1986), 77–90.

Gabbey, Alan, "Force and Inertia in the Seventeenth Century: Descartes and Newton" in Stephen Gaukroger (ed.), *Descartes, Philosophy, Mathematics, and Physics* (Harvester, 1980), 230–320.

Garber, Daniel, *Descartes's Metaphysical Physics* (University of Chicago Press, 1992).

Scott, Joseph F., *The Scientific Works of René Descartes* (Taylor and Francis, 1976).

Shea, William R., *The Magic of Numbers and Motion: The Scientific Career of René Descartes* (Science History, 1991).

Newton

General works

The best short introduction to research on Newton is:

Westfall, Richard S., "The Changing World of the Newtonian Industry," *Journal of the History of Ideas 37* (1976), 175–184.

All bibliographies on Newton, no matter how brief, must include the now standard biography of Newton:

Westfall, Richard S., *Never at Rest: A Biography of Isaac Newton* (Cambridge University Press, 1980).

Westfall has also published a much shorter and more accessible biography:

Westfall, Richard S., *The Life of Isaac Newton* (Cambridge University Press, 1993).

For a shorter but highly reliable biographical study, see:

I. B. Cohen's article on Newton in the *Dictionary of Scientific Biography,* ed. by C. C. Gillispie, vol. X (Charles Scribner's Sons, 1974), 42–103.

Another excellent biography is:

Hall, A. Rupert, Isaac Newton: *Adventurer in Thought* (Cambridge University Press, 1996).

The best collection of excerpts from Newton's writings is:

Cohen, I. B., and R. S. Westfall (eds.), *Newton: A Norton Critical Edition* (W. W. Norton, 1995).

A very useful source of information on nearly any topic related to Newton is:

Gjertsen, Derek, *The Newton Handbook* (Routledge and Kegan Paul, 1986).

The best introduction to Newton's work on mechanics in the period before 1687 is:

Wilson, Curtis, "Newton's Path to the *Principia,*" *Great Ideas Today 1985* (Encyclopædia Britannica, 1985), 179–229. This essay, which is highly recommended, draws on a much more technical essay:

_____, _____, "From Kepler's Laws, So-called, to Universal Gravitation: Empirical Factors," *Archive for History of Exact Sciences 6* (1969), 89–170.

See also:

Wilson, Curtis, "The Newtonian Achievement in Astronomy," *Planetary Astronomy from the Renaissance to the Rise of Astrophysics: Part A: Tycho Brahe to Newton,* ed. by R. Taton and C. Wilson (Cambridge University Press, 1989), 233–274.

Newton's *Principia* and Some Commentaries on It

The now standard edition of the *Principia* is:

Newton, Isaac, *The Principia: Mathematical Principles of Natural Philosophy,* trans. I. B. Cohen and Anne Whitman, preceded by "A Guide to Newton's Principia" by I. B. Cohen (University of California Press, 1999). Cohen's "Guide" (pp. 1–370) to this edition is extremely valuable.

Commentaries on the *Principia* include:

Densmore, Dana, *Newton's* Principia: *The Central Argument,* with translations and illustrations by William H. Donahue, 3rd ed. (Green Lion Press, 2003). An excellent presentation of the core argument of the *Principia.*

Brackenridge, J. Bruce, *The Key to Newton's Dynamics: The Kepler Problem and the Principia* (University of California Press, 1995). Contains a new translation of various important parts of the *Principia.*

Chandrasekhar, Subrahmanyan, *Newton's* Principia *for the Common Reader* (Clarendon Press, 1995). The author was a Nobel Prize winning astrophysicist.

Cohen, I. Bernard, *Introduction to Newton's* Principia (Harvard University Press, 1971).

Newton's Three Laws of Motion

Earman, J., and M. Friedman, "The Meaning and Status of Newton's Law of Inertia and the Nature of Gravitational Forces," *Philosophy of Science 40* (1973), 329–359.

Ellis, Brian, "The Origin and Nature of Newton's Laws of Motion" in Robert F. Colodny (ed.), *Beyond the Edge of Certainty* (Prentice Hall, 1965), 29–68.

Hankins, Thomas L., "The Reception of Newton's Second Law of Motion in the Eighteenth Century," *Archives internationale d'histoire des sciences 20* (1967), 43–65.

Hanson, Norwood Russell, "Newton's First Law: A Philosopher's Door into Natural Philosophy" in Robert F. Colodny (ed.), *Beyond the Edge of Certainty* (Prentice-Hall, 1965), 6–28.

Home, Roderick W., "The Third Law in Newton's Mechanics," *British Journal for the History of Science 4* (1968), 39–51.

Pierson, Stuart, "Corpore Cadente...: Historians Discuss Newton's Second Law," *Perspectives on Science: Historical, Philosophical, Social 1* (1993), 627–658.

_____, _____, "Two Mathematics, Two Gods: Newton and the Second Law," *Perspectives on Science: Historical, Philosophical, Social 2* (1994), 231–254.

Tajima, Nobuo, "Some Considerations on the Interpretation of Newton's Second Law and on Dijksterhuis's Opinions," *Japanese Studies in the History of Science 13* (1974), 81–86.

Newton and Philosophy

Arthur, Richard, "Space and Relativity in Newton and Leibniz," *British Journal for the Philosophy of Science 45* (1994), 219–240.

Blake, Ralph M., "Isaac Newton and the Hypothetico-Deductive Method" in Ralph M. Blake, Curt J. Ducasse, and Edward H. Madden (eds.), *Theories of Scientific Method: The Renaissance through the Nineteenth Century* (University of Washington Press, 1960), 119–143.

Butts, Robert, and John Davis (eds.), *The Methodological Heritage of Newton* (Basil Blackwell, 1970). Contains a number of relevant essays

Cohen, I. Bernard, "Hypotheses in Newton's Philosophy," *Physis 8* (1966), 163–184.

Finocchiaro, Maurice A., "Newton's Third Rule of Philosophizing: A Role for Logic in Historiography," *Isis 65* (1974), 66–73.

Harper, William, "Newton's Argument for Universal Gravitation," in I. Bernard Cohen and George Smith (eds.), *The Cambridge Companion to Newton* (Cambridge University Press, 2002), 174–201.

Lakatos, Imre, "Newton's Effect on Scientific Standards" in J. Worrall and G. Curere (eds.), *The Methodology of Scientific Research Programmes* (Cambridge University Press, 1978), 193–222.

Laymon, Ronald, "Newton's Bucket Experiment," *Journal of the History of Philosophy 16* (1978), 399–413.

McDonald, John F., "Properties and Causes: An Approach to the Problem of Hypothesis in the Scientific Methodology of Sir Isaac Newton," *Annals of Science 28* (1972), 217–233.

McGuire, J. E., "Existence, Actuality and Necessity: Newton on Space and Time," *Annals of Science 35* (1978), 463–508.

Palter, Robert, "Newton and the Inductive Method," *Texas Quarterly 10* (1967), 160–173.

Pampusch, Anita M., " 'Experimental,' 'Metaphysical,' and 'Hypothetical' Philosophy in Newtonian Methodology," *Centaurus 18* (1974), 289–300.

Popper, Karl, "The Aim of Science," *Objective Knowledge: An Evolutionary Approach* (Clarendon, 1972), 191–205.

Rogers, G. A. J., "Locke's *Essay* and Newton's *Principia*," *Journal of the History of Ideas 42* (1981), 53–72.

Smith, George W., "The Methodology of the *Principia*" in I. Bernard Cohen and George Smith (eds.), *The Cambridge Companion to Newton* (Cambridge University Press, 2002), 138–173.

Suchting, W. A., "Berkeley's Criticism of Newton on Space and Motion," *Isis 58* (1967), 186–197.

Newton and Religion

Alexander, H. G. (ed.), *The Leibniz-Clarke Correspondence* (Manchester University Press, 1956). Includes the correspondence as well as useful commentary.

Brooke, John, "The God of Isaac Newton" in John Fauvel, Raymond Flood, Michael Shortland, and Robin Wilson (eds.), *Let Newton Be!* (Oxford University Press, 1988), 169–183.

Cassirer, Ernst, "Newton and Leibniz," *Philosophical Review 52* (1943), 366–391. Deals with the differences between Newton and Leibniz and with the Clarke-Leibniz correspondence.

Cohen, I. Bernard, "Isaac Newton's *Principia,* the Scriptures, and the Divine Providence" in Sidney Morgenbesser et. al. (eds.), *Philosophy, Science, and Method: Essays in Honor of Ernest Nagel* (St. Martin's Press, 1969), 123–148.

Crowe, Michael J., *The Extraterrestrial Life Debate 1750–1900: The Idea of a Plurality of Worlds from Kant to Lowell* (Cambridge University Press, 1986), 22–25. Discusses Newton's involvement with ideas of extraterrestrial life.

Davis, Edward B., "Newton's Rejection of the 'Newtonian World View,' " *Fides et Historia 22* (1991), 6–20. Reprinted in *Science and Christian Belief 3* (1991), 103–117. Also reprinted with minor additions as "Newton's Rejection of the 'Newtonian World View': The Role of Divine Will in Newton's Natural Philosophy" in *Facets of Faith and Science,* Volume 3: *The Role of Beliefs in the Natural Sciences,* ed. by Jitse M. van der Meer (University Press of America, 1996), 75–96.

Force, James E., "Newton, the Lord God of Israel, and Knowledge of Nature" in Richard H. Popkin and Gordon M. Weiner (eds.), *Jewish Christians and Christian Jews from the Renaissance to the Enlightenment* (Kluwer Academic, 1994), 131–158.

_____, _____, "The Newtonians and Deism" in James Force and Richard Popkin (eds.), *Essays on the Context, Nature and Influence of Isaac Newton's Theology* (Kluwer, 1989), 43–73.

_____, _____, *William Whiston: Honest Newtonian* (Cambridge University Press, 1985). Discusses Whiston, Newton's successor in the Lucasian professorship who was removed from his position because of his heterodox religious views. Maintains that Whiston said what Newton thought.

_____, _____, and Richard Popkin (eds.), *Essays on the Context, Nature and Influence of Isaac Newton's Theology* (Kluwer, 1989). Contains ten essays, all by either Force or Popkin, on Newton's theological and religious views and on their influence.

Guerlac, Henry, "Theological Voluntarism and Biological Analogies in Newton's Physical Thought," *Journal of the History of Ideas 44* (1983), 219–229.

Guerlac, Henry, and M. C. Jacob, "Bentley, Newton and Providence (The Boyle Lectures once more)," *Journal of the History of Ideas 30* (1969), 307–318.

Hurlbutt, Robert H., III, *Hume, Newton, and the Design Argument* (University of Nebraska Press, 1965).

Jacob, Margaret, "Christianity and Newtonian World View" in David C. Lindberg and Ronald Numbers (eds.), *God and Nature: Historical Essays on the Encounter between Christianity and Science* (University of California Press, 1986), 238–255.

Koyré, Alexandre and I. B. Cohen, "Newton and the Leibniz-Clarke Correspondence," *Archives internationale d'histoire des sciences 15* (1962), 63–126.

Kubrin, David H., "Newton and the Cyclical Cosmos: Providence and the Mechanical Philosophy," *Journal of the History of Ideas 28* (1967), 325–346. An important study regarding Newton and religion.

Manuel, Frank, *The Religion of Isaac Newton* (Clarendon, 1974).

McLachlan, Herbert, *Sir Isaac Newton: Theological Manuscripts* (Liverpool University Press, 1950). After an introduction, the editor presents some previously unpublished theological manuscripts by Newton.

Perl, Margula, "Physics and Metaphysics in Newton, Leibniz, and Clarke," *Journal of the History of Ideas 30* (1969), 507–526.

Pfizenmaier, Thomas, "Was Isaac Newton an Arian?" *Journal of the History of Ideas 58* (1997), 57–80.

Priestley, F. E. L., "The Clarke-Leibniz Controversy" in Robert Butts and John Davis (eds.), *The Methodological Heritage of Newton* (Basil Blackwell, 1970), 34–56.

Schaffer, Simon, "Newton's Comets and the Transformation of Astrology" in Patrick Curry (ed.), *Astrology, Science and Society: Historical Essays* (Boydell, 1987), 219–243.

Stewart, Larry, "Samuel Clarke, Newtonianism, and the Factions of Post-Revolutionary England," *Journal of the History of Ideas 42* (1981), 53–72.

_____, _____, "Seeing through the Scholium: Religion and Reading Newton in the Eighteenth Century," *History of Science 34* (1996), 123–165.

Strong, E. W., "Newton and God," *Journal of the History of Ideas 13* (1952), 147–167.

Trengove, Leonard, "Newton's Theological Views," *Annals of Science 22* (1966), 277–294.

Vailati, Ezio, *Leibniz & Clarke: A Study of Their Correspondence* (Oxford University Press, 1997).

Verlet, Loup, " 'F = ma' and the Newtonian Revolution: An Exit from Religion through Religion," *History of Science 34* (1996), 303–346.

Wagner, Fritz, "Church History and Secular History as Reflected by Newton and His Time," *History and Theory 8* (1969), 97–111.

Westfall, Richard S., "Isaac Newton" in *The History of Science and Religion in the Western Tradition: An Encyclopedia,* ed. by Gary Ferngren (Garland, 2000), 95–99.

_____, _____, "Isaac Newton's 'Theologiae Gentilis Origines Philosophiae,' " in W. Warren Wager (ed.) *The Secular Mind: Transformation of Faith in Modern Europe* (Holmes & Meier, 1982), 15–34.

_____, _____, "Newton's Theological Manuscripts" in Zev Bechler (ed.), *Contemporary Newtonian Research* (Reidel, 1982), 129–143.

Westfall, Richard S., "The Rise of Science and the Decline of Orthodox Christianity: A Study of Kepler, Descartes, and Newton" in David C. Lindberg and Ronald Numbers (eds.), *God and Nature: Historical Essays on the Encounter between Christianity and Science* (University of California Press, 1986), 218–237.

_____, _____, *Never at Rest: A Biography of Isaac Newton* (Cambridge University Press, 1980). See, for example, the treatment of Newton's General Scholium, 744–745, 748–749; for bibliography, see note 58 on p. 749; for the Leibniz-Clarke correspondence, see pp. 777–780.

Newton and Alchemy

Dobbs, Betty Jo Teeter, *The Foundations of Newton's Alchemy or "The Hunting of the Green Lyon"* (Cambridge University Press, 1975), which is a standard source on Newton's work in alchemy.

Hall, Marie Boas, "Newton's Voyage on the Strange Seas of Alchemy" in M. L. Righini Bonelli and W. R. Shea (eds.), *Reason, Experiment and Mysticism in the Scientific Revolution* (Science History Publications, 1975), 239–246.

Rattansi, Pyarali M., "Newton's Alchemical Studies" in Allen G. Debus (ed.), *Science, Medicine and Society in the Renaissance,* vol. 2 (Science History Publications, 1972), 167–182.

Westfall, Richard S., "Newton and Alchemy" in Brian Vickers (ed.), *Occult and Scientific Mentalities in the Renaissance* (Cambridge University Press, 1984), 315–335.

_____, _____, "The Influence of Alchemy on Newton" in Marsha P. Hanen (et al., eds.), *Science, Pseudo-Science, and Society* (Wilfrid Laurier University Press for the Calgary Institute for the Humanities, 1980), 145–169.

_____, _____, "The Role of Alchemy in Newton's Career" in M. L. Righini Bonelli and W. R. Shea (eds.), *Reason, Experiment, and Mysticism in the Scientific Revolution* (Science History Publications, 1975).

Other Valuable Studies Relevant to Newton

Christianson, Gale E., *In the Presence of the Creator: Isaac Newton and His Times* (Free Press, 1984).

Cohen, I. Bernard, "Newton and Keplerian Inertia: An Echo of Newton's Controversy with Leibniz" in Allen G. Debus (ed.), *Science, Medicine and Society in the Renaissance,* vol. 2 (Science History, 1972), 199–211.

_____, _____, "Newton's Discovery of Gravity," *Scientific American 244* (March, 1981), 166–179.

_____, _____, "Quantum in se est," *Notes and Records of the Royal Society 19* (1964), 131–155. Newton's concept of inertia in relation to Descartes and Lucretius.

_____, _____, "The Thrice Revealed Newton" in Trevor H. Levere (ed.), *Editing Texts in the History of Science and Medicine* (Garland, 1982), 117–184. Strikingly shows the difference between the image of Newton in the period after his death and the image that has emerged from recent scholarly studies on Newton's letters and manuscripts.

_____, _____, *The Newtonian Revolution* (Cambridge University Press, 1980).

Densmore, Dana, "At What Point in *Principia* Does Newton Believe Universal Gravitation Is Established?" in *Essays in Honor of Curtis Wilson* (MIT Press, forthcoming).

Fauvel, John, Raymond Flood, Michael Shortland, and Robin Wilson (eds.), *Let Newton Be!* (Oxford University Press, 1988). Contains accessible and authoritative essays on many aspects of Newton's life and work.

Gingerich, Owen, "Circumventing Newton: A Study in Scientific Creativity," *American Journal of Physics 46* (1978), 202–206.

Hall, A. Rupert, "Huygens and Newton" in *The Anglo-Dutch Contribution to the Civilization of Early Modern Society: An Anglo-Netherlands Symposium* (Oxford University Press for the British Academy, 1976), 45–59.

___, _____, *All Was Light: An Introduction to Newton's* Opticks (Clarendon Press, 1993).

___, _____, *Philosophers at War: The Quarrel between Newton and Leibniz* (Cambridge University Press, 1980). Discusses the dispute between Newton and Leibniz concerning priority in the invention of the calculus.

Herivel, John, *The Background to Newton's* Principia. *A Study of Newton's Dynamical Researches in the Years 1664–84* (Clarendon Press, 1965).

Hoskin, M. A., "Newton, Providence and the Universe of Stars," *Journal for the History of Astronomy 8* (1977), 77–101.

Iltis, Carolyn, "The Leibnizian-Newtonian Debates: Natural Philosophy and Social Psychology," *British Journal for the History of Science 6* (1973), 343–377.

Jacob, M. C., and Henry Guerlac, "Bentley, Newton and Providence (the Boyle Lectures Once More)," *Journal of the History of Ideas 30* (1969), 307–318.

Keynes, J. M., "Newton, the Man" in Royal Society, *Newton Tercentenary Celebrations* (Cambridge University Press, 1947), 27–34.

Keynes, Milo, "The Personality of Isaac Newton," *Notes and Records of the Royal Society 49* (1995), 1–56. Presents an effective survey of this topic as well as of the personal life of Newton. Written by a nephew of J. M. Keynes.

Koyré, Alexandre, *From the Closed World to the Infinite Universe* (Johns Hopkins University Press, 1957). A classic study by a scholar who in this and other works shed much light on the scientific revolution and on Newton.

_____, _____, *Newtonian Studies* (Harvard University Press, 1965).

_____, _____, and I. B. Cohen, "Newton and the Leibniz-Clarke Correspondence, with Notes on Newton, Conti and des Maizeaux," *Archives internationales d'histoire des sciences 15* (1962), 63–126.

_____, _____, and I. B. Cohen, "The Case of the Missing Tanquam: Leibniz, Newton and Clarke," *Isis 52* (1961), 555–566.

Kubrin, David, "Newton's Inside Out!: Magic, Class Struggle, and the Rise of Mechanism in the West" in Harry Woolf (ed.), *The Analytic Spirit: Essays in the History of Science in Honor of Henry Guerlac* (Cornell University Press, 1981), 96–121.

Laymon, Ronald, "Newton's Demonstration of Universal Gravitation and Philosophical Theories of Confirmation" in John Earman (ed.), *Testing Scientific Theories* (University of Minnesota Press, 1983), 179–199.

Manuel, Frank, *A Portrait of Isaac Newton* (Frederick Muller, 1980).

Martins, Roberto De A., "Huygens's Reaction to Newton's Gravitational Theory" in J. V. Field and Frank A. J. L. James (eds.), *Renaissance and Revolution: Humanists, Scholars, Craftsmen and Natural Philosophers in Early Modern Europe* (Cambridge University Press, 1993), 203–213.

McGuire, J. E., "Atoms and the 'Analogy of Nature': Newton's Third Rule of Philosophizing," *Studies in the History and Philosophy of Science 1* (1970), 3–58.

McKie, D. and G. de Beer, "Newton's Apple," *Notes and Records of the Royal Society 9* (1952), 46–54, 333–335.

Nauenberg, Michael, "Hooke, Orbital Motion, and Newton's *Principia*," *American Journal of Physics 62* (1994), 331–350.

Palter, R. (ed.), *The Annus Mirabilis of Sir Isaac Newton 1666–1966* (MIT Press, 1970). First appeared as vol. 10, no. 3 (1967) of the *Texas Quarterly.*

Pourciau, Bruce, "Reading the Master: Newton and the Birth of Celestial Mechanics," *American Mathematical Monthly 104* (1997), 1–19.

Rickey, V. Frederick, "Isaac Newton: Man, Myth, and Mathematics," *College Mathematics Journal 18* (1987), 362–389. A reliable and accessible introduction to Newton's work, especially in mathematics.

Russell, J. L., "Kepler's Laws of Planetary Motion: 1609–1666," *British Journal for the History of Science 2* (1974), 1–24.

Sailor, D. B., "Moses and Atomism," *Journal of the History of Ideas 25* (1964), 3–16.

Toulmin, Stephen , "Science, Philosophy of," *Encyclopædia Britannica,* vol. 16, 15th ed. (Encyclopædia Britannica, Inc., 1983), pp. 375–393.

Wallis, Peter, and Ruth Wallis, *Newton and Newtoniana 1672–1975* (Dawson, 1977). Consists of a bibliography, arranged by subject, of publications by and about Newton.

Westfall, Richard S., "Newton and the Fudge Factor," *Science 179* (1973), 751–758.

————, ————, "Newton's Marvelous Years of Discovery and Their Aftermath: Myth Versus Manuscript," *Isis 71* (1980), 109–121.

————, ————, *Science and Religion in Seventeenth-Century England* (Yale University Press, 1958).

Whiteside, Derek T., "Before the *Principia*: The Maturing of Newton's Thoughts on Dynamical Astronomy," *Journal for the History of Astronomy 1* (1970), 5–19. Whiteside is the leading authority of Newton's mathematical writings.

————, ————, "Newton's Marvellous Year: 1666 and All That," *Notes and Records of the Royal Society 20* (1966), 32–41.

————, ————, "Newton's Thoughts on Planetary Motion: A Fresh Look," *British Journal for the History of Science 2* (1964), 117–138.

————, ————, "The Prehistory of the *Principia* from 1664 to 1686," *Notes and Records of the Royal Society 25* (1991), 11–61.

Wilson, Curtis, "Newton and Some Philosophers on Kepler's 'Laws,'" *Journal of the History of Ideas 35* (1974), 231–258.

Yoder, Joella, *Unrolling Time: Christiaan Huygens and the Mathematization of Nature* (Cambridge University Press, 1988).

Newtonian Sites and Memorabilia, etc.

Adrian, Lord, "Newton's Rooms in Trinity," *Notes and Records of the Royal Society 18* (1963), 17–24.

Haskell, Francis, "The Apotheosis of Newton in Art" in Robert Palter (ed.), *The Annus mirabilis of Sir Isaac Newton 1666–1966* (MIT Press, 1970), 302–321.

Richardson, A. E., "Woolsthorpe Manor House," *Notes and Records of the Royal Society 1* (1947), 34–35.

Smith, D. E., "Portraits of Sir Isaac Newton" in W. J. Greenstreet, *Isaac Newton (1642–1727)* (G. Bell and Sons, 1927), 171–178.

Some Disciples or Opponents of Newton, the Period after Newton, and the Influence of Newton

Arons, A. B., "Newton and the American Political Tradition," *American Journal of Physics 43* (1975), 209–213.

Boss, V. I., *Newton and Russia, the Early Influence 1698–1796* (Harvard University Press, 1972).

Cohen, I. Bernard, "The French Translation of Isaac Newton's *Philosophiae naturalis principia mathematica* (1756, 1759, 1766)," *Archives internationales d'histoire des sciences 72* (1969), 37–67.

Ferguson, J., *Dr. Samuel Clark* (Roundwood Press, 1976).

Guerlac, Henry, *Newton on the Continent* (Cornell University Press, 1981).

Hetherington, Norriss S., "Isaac Newton's Influence on Adam Smith's Natural Laws in Economics," *Journal of the History of Ideas 44* (1983), 497–505.

Nicolson, M. H., *Newton Demands the Muse* (Princeton University Press, 1966). A study of how Newton's writings, especially his *Opticks,* influenced literature.

Schneider, Michael, "Of Mystics and Mechanism: The Reception of Newtonian Physics in 18th-Century English Poetry," *Synthesis* (Cambridge, MA), *1 (1)* (1972), 14–26.

Staum, Martin S., "Newton and Voltaire: Constructive Skeptics," *Studies on Voltaire and the Eighteenth Century 62* (1968), 29–56.

Stewart, Larry, "Samuel Clarke, Newtonianism, and the Factions of Post-Revolutionary England," *Journal of the History of Ideas 42* (1981), 53–72.

Taton, René, "Madame du Chatelet, traductrice de Newton," *Archives Internationale d'Histoire des Sciences 22* (1969), 185–210.

Walters, R. L., "Voltaire, Newton and the Reading Public" in Paul Fritz and David Williams (eds.), *The Triumph of Culture: 18th-century Perspectives* (Hakkert, 1972), 133–155.

Mechanics between Newton and Einstein

Hankins, Thomas L., *Science and the Enlightenment* (Cambridge University Press, 1985).

Harman, Peter M., *Energy, Force and Matter: The Conceptual Development of Nineteenth-Century Physics* (Cambridge University Press, 1982).

Holton, Gerald, *Introduction to Concepts and Theories in Physical Science,* revised by Stephen G. Brush (Princeton University Press, 1985). Although designed primarily as a physics textbook, it is rich in historical information.

Nye, Mary Jo (ed.), *Modern Physical and Mathematical Sciences* (Cambridge University Press, 2002).

Purrington, Robert D., *Physics in the Nineteenth Century* (Rutgers University Press, 1997).

Einstein

Bernstein, Jeremy, *Einstein* (Viking, 1972). A good elementary presentation of Einstein's life and thought.

Clark, Ronald W., *Einstein: The Life and Times* (World, 1971). A biography of Einstein written on a popular level.

Einstein, Albert, and Leopold Infeld, *The Evolution of Physics* (Simon and Schuster, 1961). Written in 1938, this book contains an elementary exposition, without the use of equations, of both the special and general theories as well as of quantum theory, to which Einstein also made crucial contributions. Not reliable for historical information.

_____, _____, "Autobiography" in Paul Arthur Schilpp (ed.), *Albert Einstein: Philosopher-Scientist,* vol. I (Harper and Row, 1959), 1–95. Written when Einstein was 67, this work is highly interesting if at times somewhat technical. Actually it is only 42 pages long, the German original having been presented on facing pages.

_____, _____, H. A. Lorentz, H. Weyl, and H. Minkowski, *The Principle of Relativity* (Dover, n.d.). A collection of papers by Einstein and other physicists. Among the items included are translations of Einstein's 1905 special theory of relativity paper and his 1916 general theory of relativity paper.

_____, _____, *Relativity: The Special and General Theory,* trans. R. W. Lawson (Crown, 1961). An accessible presentation on which the presentation in this work is partly based.

French, A. P. (ed.), *Einstein: A Centenary Volume* (Harvard University Press, 1979). An anthology of generally fine essays on Einstein and his ideas.

_____, ____, *Special Relativity* (W. W. Norton, 1968).

Goldberg, Stanley, *Understanding Relativity* (Birkhäuser, 1984). A fine presentation written from a historical point of view.

Highfield, Roger, and Paul Carter, *The Private Lives of Albert Einstein* (St. Martin's Press, 1993). A well researched volume drawing on Einstein's correspondence, some of which is only now becoming known.

Holton, Gerald, "Einstein, Michelson, and the Crucial Experiment," *Isis 60* (1969), 122–197. Holton shows in this paper that the Michelson-Morley experiment did not play a significant role in Einstein's formulation of the special theory of relativity.

Katz, Robert, *An Introduction to the Special Theory of Relativity* (Van Nostrand, 1964).

Kragh, Helge, *Quantum Generations: A History of Physics in the Twentieth Century* (Princeton University Press, 1999).

Miller, Arthur I., *Albert Einstein's Special Theory of Relativity: Emergence (1905) and Early Interpretation (1905–1911)* (Addison-Wesley, 1981). A very useful source for the early history of the special theory.

Pais, Abraham, *'Subtle Is the Lord': The Science and the Life of Albert Einstein* (Clarendon, 1982). An excellent high level biography of Einstein.

Ray, Christopher, *The Evolution of Relativity* (Adam Hilger, 1987).

Rigden, John, *Einstein 1905: The Standard of Greatness* (Harvard University Press, 2005). An accessible presentation of the key discoveries Einstein published in 1905.

Segré, Emilio, *From Falling Bodies to Radio Waves: Classical Physicists and Their Discoveries* (Freeman, 1984). A good elementary presentation of the developments in physics in the three centuries before Einstein.

Williams, L. P., *Relativity Theory: Its Origins & Impact on Modern Thought* (John Wiley, 1968).

Index

Photos and graphics information

Portrait of Galileo (p. 28): Chalk drawing by Ottavio Leoni, 1624. Biblioteca Marucelliana, Florence, Italy. By permission of Scala / Art Resource, NY.

Portrait of Newton (p. 110): Mezzotint by MacArdel after a painting by E. Seeman, 1726. From *Sir Isaac Newton 1727–1927. A Bicentenary Evaluation of His Work* (History of Science Society Publications, 1928).

Photograph of Einstein (p. 268): Yousuf Karsh, 1948.

Cover illustration by William H. Donahue based on an engraving in Isaac Newton, *System of the World* (1728), with a cannon from Galileo, *Two New Sciences*, and images of Sputnik and the Apollo command module.

Line drawings by William H. Donahue.